A
TACHYON
THEORY
OF
EVERYTHING

A
TACHYON
THEORY
OF
EVERYTHING

Harold Arnold
Process Engineer
Teledyne S&I

To

Wolfgang Oswald Joseph Arnold

My dad knows everything!

Preface

When I was 16 or 17 years old we saw (One TV in the house and dad always prioritized educational specials over anything else.) a couple of specials on Einstein within a couple of weeks of each other. I think the repetition helped stick the ideas in my otherwise distracted head, especially the "stretchy shrinky space" ,as I tried to explain to my friends at the time. I was also amused by the notion that Einstein was a distracted teen just like me, and therefore I might have as good a chance of coming up with something new as he did.

So, endeavoring to think sincerely about stretchy shrinky space, I came up with the vague notion that there is a type of space "outside of time" where distance doesn't matter. It takes the same time to travel between any two points regardless of the measurable distance between them. Then, filling such a space with a uniform grid of points, I can imagine a Euclidean type space. Because now to go to the point twice as far, I have to go through twice as many points taking twice the time. Due to the Beatles song I was listening to when I thought of this, I called this space the Void, still, vague notions, cloudy visions.

I was introduced to tachyons in my Special Relativity class by Dr Saenz at Cal Poly S.L.O. in the mid 80s . It was in a side note, nothing in the text. He mentioned the bits about being possible since they aren't explicitly forbidden, and their being everywhere in their trajectory at once, and the inverse energy-velocity relation along with the Cherenkov paradox for charged tachyons. As he concluded the lecture with the note that nobody gets a PhD for a negative result (Which is tragic.), I was looking around the room to see if anybody else saw what I was looking at, and that was that these were to very particles of space-time, but they all looked like they always did. I was mentally in orbit, I knew these were the answer.

Realizing in grad school that I lacked the math to explore physics the way I wanted to (I actually probably lacked the confidence more than the ability.), I dropped out, and took math classes after work to prepare to work on my own. I was in a Real Analysis class at Cal State Dominguez Hills in the mid 90s where I was introduced to the metric space that describes my void.

In the late 90s I decided to pursue this theory in earnest and started reviewing my old physics texts, I realized I still wanted to take a General Relativity class to verify my thoughts on gravity and I started going back and fourth in my Quantum Mechanics book trying to make sense of it.

Around '07 (On the verge of giving up, I knew you weren't supposed to make sense of QM.) at Rockwell in Camarillo, CA, when explaining why I was asking for help with QM problems, and I was in-articulately trying to explain my theory, Dr. Mason Thomas, instead of dismissing my nonsense, quite sincerely suggested I write things down to organize and clarify my thoughts. So I started writing. Then a few weeks later I was discussing Brian Greens book about String Theory with Dr. Donald Lee, who was also supportive and engaged in arguments about what might be wrong with string theory, but more importantly, what are the properties that would make any theory attractive. Not long after that I came up with the whole Available State – Uncertainty Principle – Single Photon Interference concept, and I knew I was onto something and I would be at this the rest of my life (That's right, I'm implying that my notions make sense of QM!). The notion seemed consistent enough that I put it aside to fit the rest of physics together.

A couple of years later I randomly walked into the notation $v = (c, U)$, and started to develop my current mental model of Shape. I played a little bit, again, just to satisfy myself that I was on the right track.

Then I studied enough General Relativity to satisfy myself that my thoughts on gravity were still good.

The final mystery was Electricity and Magnetism. It took a few years before I finally added the helical action. It was such a contrived idea that I decided I had to test this idea in some mathematical detail before doing anything else including working the idea back through the rest of the theory.

The electric part was easy, but the magnetism, again took some time. Working out the 3-D vector equations for the force on a charge in an EM field, I was again satisfied that the helix was a good idea, so I worked the idea throughout the rest of the theory and found everything fitting together like that final piece of a puzzle.

I think things fit so well and the idea so good, that I think its publishable, so I gave myself a dead line to come up with some low hanging fruit in the Shape department. Realizing that deriving the masses of the particles was going to be another several year effort, I have cinched things up and present my theory here.

I do not mathematically prove things. I try to paint a picture of a model with a list of notions for how the model explains the various properties of physics. Each notion is only a paragraph or so long, some clearer than others. This hopefully provides a context in which to tie the separate branches of physics together, and also give coherent mental pictures to things that previously had no mental picture, or disparate unconnected pictures.

The hope is that folks with this idea have a new perspective with witch to think about physics.

Table of Contents

Table of Contents

Table of Contents

Table of Contents

Table of Contents

A
TACHYON
THEORY
OF
EVERYTHING

Introduction

Introduction: What are tachyons?

Start with the usual Special Relativity equations;
E = energy, m = mass, c = speed of light, v = velocity

$$E = mc^2 \text{ , } \beta = \frac{v}{c} \text{ , } m = \frac{m_0}{\sqrt{1-\beta^2}} \text{ , } m_0 = \text{rest mass, } E = \frac{m_0 c^2}{\sqrt{1-\beta^2}} \text{ .}$$

Then solving for the speed, β , results in

$$\beta = \sqrt{1 - \frac{\left(m_0 c^2\right)^2}{E^2}}$$

Plotted below for three different types of rest mass.

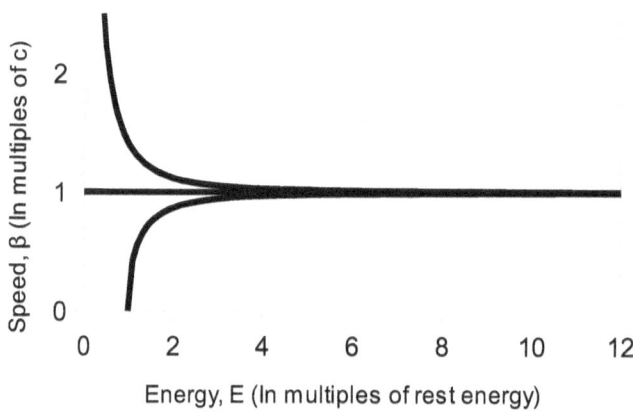

3 Cases to get a Real speed, β ;
    1) Tardyons (That's what Dr Saenz called them, Not bradyons.) (Bottom curve): The rest mass is real and greater than zero. Total energy is equal to or greater than the rest energy. The velocity is less than c, and asymptotically approaches c as the energy goes to infinity.
    2) Luxons (Horizontal line at $\beta = 1$): The rest mass is zero.

A Tachyon Theory of Everything

Energy is no longer part of the equation. The velocity is c no matter what the energy is.

    3) Tachyons (Top curve): The rest mass is Complex $(m_o = \pm im_R, i = \sqrt{-1}$ where m_R is real), so

$$\beta = \sqrt{1 - \frac{\left(im_R c^2\right)^2}{E^2}} = \sqrt{1 + \frac{m_R^2 c^4}{E^2}}$$

The square root is never negative. As E goes to zero, β goes to infinity, while as E goes to infinity, the velocity asymptotically approaches c. No guarantee that m_o is constant, or E is greater than its rest energy.

    Also, looking at the relativistic time equation;

$$\Delta t \sqrt{1 - \beta^2} = \Delta t', \text{ for } \beta > 1, i\Delta t \sqrt{1 - \beta^2} = \Delta t', \text{ and for } \beta \gg 1, i\Delta t \beta = \Delta t'$$

For v < c, t' is real
For v = c, t' is zero
For v > c, t' is complex
    So,
For v < c, the particle has a real rest mass and travels in a real time
For v = c, the particle has zero rest mass and stands still in time.
For v > c, the particle has a complex rest mass and travels in a complex time

    I need two types of tachyon for the model, $\pm im$ with $\pm it$ gives four.. I will label the two "$\pm im$" for now.

    Some rationalizations for believing in tachyons. Both their complex and faster than light nature;
    That photons are mass less and standing still in time seems no more or less intuitive than a complex mass in a complex time (to me anyway). The first two of the three cases, slower than light

real masses and equal to light zero rest massed photons, exist in abundance. Thus I think if tachyons exist, they're just as abundant as the first two cases. That they haven't been discovered yet leads me to believe that there is an interpretational issue. That is, they are all around, but we aren't recognizing them as tachyons.

As shown in reference 7, Complex values can be used to represent extended particles in Quantum Mechanics (QM).

The instantaneous change of a QM wave function throughout space hints that something is traveling faster than light.

Invariant charge on an electron tells me that it must be conveyed by something going much faster that light or it would Doppler shift and an approaching charge would have a larger charge than a receding one, but they look the same. Length contraction makes the field normal to travel stronger, but head on versus directly receding look the same.

How else could a Black Hole have a charge if it's not conveyed by something going faster than c?Allowing virtual photons to violate physics beyond the uncertainty principle is just as big a stretch as some of my logical contortions in this theory.

Likewise for gravitons and Black Holes. Since gravitons theoretically travel at c, aren't they also trapped behind the Event Horizon? Or do we allow real gravitons to convey the gravity of normal masses, but revert to virtual gravitons for the conveyance of a Back Hole's gravity?

Then there is the Inflationary theory that has the early universe expanding much faster than c.

Finally, Michelson-Morley disproved the existence of a Tardyonic Ether, not a Tachyonic one.

A Tachyon Theory of Everything

Where to begin? When I first started writing I had the notions of the Void, the tachyon being everywhere at once so it can self-interfer to create a real mass, and that they are tachyonic with respect to each other, but not much else. So I wrote the following Symmetry Table to try to inspire some thoughts.

Since Relativity says that objects traveling at different speeds are also traveling at different times, I will use the terminology of time when addressing issues of velocity to take advantage of space-time semantics.

Symmetry Table

TARDYONS	TACHYONS
Symmetry 1	
Spatially displace each other.	Can all share the same space
Can all share the same velocity (time)	Temporally displace each other
Symmetry 2	
Spatially extant (or variable)	Spatial points (or a single constant size) with a minimum velocity in space.
Temporal points (or a single constant size in time) with a minimum velocity in time.	Temporally extant (or variable)
Symmetry 3	
3-D Space, 1-D time, 1-D mass	1-D space, 3-D time, 3-D mass

When I explained electricity with a universally constant tachyon size (and thus eliminating point particles), I modified some table entries with,"or a single constant size ".

Where to begin

What the above makes me think about is the following;

TIME is Complex, Extant and Multidimensional

    Complex time allows for temporal interference. By being in Complex time and thus orthogonal to Real time, I can say that the tachyon exists forward and backward in our time at the same time. This fits with being everywhere in it's trajectory at once. With this I can allow the tachyon to temporally interfere with itself. IF it can return upon itself (within it's own light cone), then it can interfere with itself to create a real mass by multiplying together two interfering complex masses. Some different ways to do this include rationalizing the back interfering with the front to create a -m, then two -m's interfere to create a +m (resulting in space being filled up with -m?), or rationalize the returning one to be the complex conjugate of the one being returned upon (maybe the front of the tachyon has a +it and the back has a -it). Also note that although the complex nature has been multiplied out, I want the real mass to still maintain it's "+" or "-" nature for collision purposes. Because the tachyon is a particle with a velocity, I will be treating it as a particle sometimes, and a line at others.

    Temporal extent means rotating and looping. Using the equivalence of velocity and time, a spinning object has different parts moving at different velocities. It has an extension of velocities, or an extension of time frames. If a tachyon, with it's minimum velocity has orient-able sides that are traveling at different velocities, then clearly circular patterns can be swept out allowing for the self interference mentioned above. Considering Axiom 1 with the tachyons rotating and twisting to get out of each other's light cones, These orient-able sides will determine the directions of the rotating and twisting. The size of the loop can be anything from microscopic to the size of the universe. I need the vast majority of loops to be big enough to be locally straight lines

5

when viewed on the scale of whatever maximum distance the inverse r squared law has been observed for gravity and electricity (Minimum loop size about that of a typical galaxy?)

So I've got a tachyon traveling in a giant loop and returning upon itself to produce a tiny point of real mass

More Dimensions/Extensions in time to spin the mass. Symmetry 3 suggests more interesting patterns can be swept out to include possible spins, precessions, etc. One possible thought is to take the circle described above and, at the point of self interference, rotate it to sweep out a torus, then with another dimension, precess the torus, etc.

I want the "handedness" of the spins of the "+" and "–" complex masses to be opposite

Mass is also Complex, Extant, and Multidimensional

For the **Complex** aspect of tachyons, I wish to consider them as complex by themselves and not to be squared or conjugated before considering their nature. I will get the square or conjugate from the self interference mentioned in symmetry 1 above with the ability to occupy the same space at the same time and thus interfere (the interference being the rationalization to square or conjugate.) to produce a real mass. I wish to look at the notion of complex being orthogonal to real in a physical sense and ask what is an orthogonal property to real mass? Well, one simple property of mass is that you can bump into it. So I will assert that the complex nature of tachyon mass means that you do not bump into it. Just like you don't bump into space.

Extended mass means different sides having different masses. And might mean collisions on different sides see different masses thus resulting in a curved path in an otherwise uniform cloud of collisions.

Where to begin

The **Multidimensional** nature was implied with the complex time. The minimum velocity defines a front and back, while the looping action defines an inside and outside which then defines the left and right. Each side traveling at a different velocity and thus having a different mass.

So, multidimensional, extended, complex time and mass produce inherent angular velocities and temporal and mass interferences that work together to create a real mass in a real time from a complex mass in a complex time, and spin it. This is similar (semantically anyway) to Einstein trying to show that matter was knots in time. (Actually, I'll consider particles to be photons tied in knots)

The above was my first grand intuitive notion of what might be. I haven't thought about parts of it beyond writing it down, but these notions have remained in the back of my mind as I have fleshed out a model with enough structure to provide feed back with ideas of what further properties might be needed to explain physics.

A Tachyon Theory of Everything

The Tachyon Properties

Axiom 0, Tachyons can be thought of as lines. This has already been stated, I forgot who said it, but he said (I paraphrase), "A tachyon, being faster than light, is everywhere in it's trajectory at once.". I will use this to rationalize Temporal Interference, the creation of real mass.

Axiom 1, Tachyons are tachyonic w.r.t. each other. If two tachyons, within each other's light cones, have velocities within c of each other, they repel each other to be outside, or at least on the edge of, each other's light cones. The three ways I can imagine doing this are

1) Speed adjustment – If one tachyon appears near, parallel, and at the same speed as a second tachyon, the second tachyon gets an acceleration (plus or minus) of c along it's current trajectory.(I have rejected this one and no longer consider it.)

2) Push Away – Now the second tachyon gets pushed directly away from the first tachyon.(I use this one to produce photons and to give energy to decay products.). And

3) Rotate – The two parallel tachyons rotate and twist their velocities to be c apart. (I use this one to get the tachyons traveling in giant loops to come back upon themselves.).

Helical Trajectories. On top of the looping and spinning mentioned above, I wish to impose a helical path. So now the simple circle/loop mentioned earlier is replaced by a very long slinky formed into a circle/loop. I will split the total tachyon energy into the rotational kinetic energy of traveling around the slinky spirals and the linear kinetic energy of traveling along the loop. I will usually distinguish the parts of the curve making up the trajectory as the loop and helix, although the path of the loop may be a helix itself. +im tachyons will have clock wise helical action while -im tachyons will have the opposite.

Mass Phase: Right now I imagine that the ultimate self interference resulting in a scalar mass has to happen with the two interfering masses oriented just right. This keeps randomly crossing tachyons from creating real masses, but also creates a "Phase Constraint" on any interactions the tachyon may encounter around it's loop. The orbit must be changed such that the self interference still occurs.

To keep the tachyon's Real Particle's mass the same, the total velocity must be constant through any interaction. To keep the orientations at the Real particle the same, the radius of the small helix must adjust with the velocities.

Particles: I have very very tiny complex tachyon particles, not points, creating very very tiny real spherical masses and sweeping out very large, galactic sized trajectories/paths. If complex time and mass are bending the paths according to definite rules, then the tachyons are bent into definite shapes. Different particles will simply be different shapes. But the shape is probably not as important as the energy, and angular and linear momenta that are generating it. My usual mental vision of a simple, but galactic sized circle or figure 8, sweeping out a torus, with the electron as the tiny sphere of real mass filling in the point sized doughnut hole of the torus is probably wrong in the details, but conveys the basic idea.

Vectors will be denoted with bold when in type font, and an arrow over the variable when in math font.

A Tachyon Theory of Everything

Collisions/Interactions

Tachyon-Tachyon Interaction: Considering Axiom 1 with tachyons being tachyonic w.r.t. each other; When **two tachyons** are traveling within c of each other, they are repelled ($1/r^2$ dependence?). I will call this a "Temporal Interaction/Collision".

When **two centers of Complex mass** are traveling within c of each other they are repelled ($1/r^2$ dependence?). This is also called a Temporal Interaction or collision.

Note that centers of real masses are not repelled. The three centers of complex mass are
1) the tachyon itself, 2) the loop's center, and 3) the helix's center.

Repulsion might also be proportional to the energies/masses being repelled. Thought I needed this for gravity, but the notion, developed in the Shape appendix that the loop radius is inversely proportional to the energy may negate the need for this.

Don't know if a tachyon-tachyon spatial collision/intersection needs any kind of interaction.

Temporal Pressure: A cloud of tachyons repelling each other as stated will result in the summed repulsions creating a pressure.

As understood from the first page, the velocity determines the mass/energy BUT temporal pressure also modulates the mass/energy. So I allow the rest mass of the tachyon to be modulated by the temporal pressure.

Tachyon-Real Mass Interaction: When a tachyon passes through a real mass, depending on the sign of both the real mass and the tachyon's, the tachyon exchanges some of it's rotational kinetic energy (KE) for it's linear KE but keeps its total mass energy unchanged,

For starters let +im tachyons have a + angular momentum vector (+**L**) (Bold indicates vector.) to their helical action. That is the angular momentum vector, **L**, in a right handed coordinate system is pointing in the same direction as its loop velocity whereas the -im tachyon's helical angular momentum vector is pointing in the opposite direction (-**L**) as it's travel along the loop.

When a +im tachyon goes through a +im's real mass, it gains a bit of rotational KE and looses a bit of linear KE. While going through a -im's real mass results in loosing a bit of rotational KE while gaining in the linear KE. Then visa versa for -im tachyons going through the real masses.

For example, the real mass of an electron, being made up of a single -im tachyon, has all the -im tachyons of the cloud passing through it increase in there -**L** while all of the +im tachyons have their +**L** decreased, so what was a neutral space without the charge present is now a gradient of net -**L** emanating from the charge. This will be used to explain electric charge and the electric field.

Also, the delta **L** is greater for those cloud tachyons moving and spinning with the real mass. But the maximum Δ**L** is universally constant for all tachyons (h bar?).

Finally, the Tachyon-Real Mass interaction involves what I call a **time "tic"**. An advancement of time. This unit of time is shorter than anything yet measured, a shortest possible time.

Real Mass-Real Mass Interaction: Two Real masses (particles or photons) interacting takes time. A **time delay** keeps the particles at the same points for multiple time tics.

A Tachyon Theory of Everything

The Void: There is a metric space, the "Discrete Metric" as Wikipedia calls it, where the distance between any two different points is 1 regardless of the points' "Euclidean Coordinates". This space has 3 space dimensions, but no time. It takes the same time and effort (energy?) to go between any two points regardless of their coordinates. **Time tics at the points**. Using a uniform cloud of tachyons to define a uniform "grid" of points, I can see a uniform velocity c, and a flat euclidean space. But where the points are denser, the spacing has "shrunk" and visa versa. In empty neutral space, the tachyons are uniformly distributed in the "flat" cloud. So the Void is divided up by tachyons providing points in space and tics in time. **A tachyon is an element of space-time**.

Available States: Since tachyons displace each other (temporally, as described above under interactions), they can fill up space and therefore leave or create available states. These states flow about the cloud as waves. The entire available state of the tachyon (and the particle it's making) probably has real and complex parts. Two (or more?) available states may share two (or more?) tachyons (and the particles they're making) to be a **Shared Available State.**

The tachyon's (and the particle it's making's) properties of mass/energy are determined by it's available state, while position and time are determined by the tachyon (and the particle it's making) itself.

Probably Proximate Property (PPP): Surrounding the available state containing a tachyon of interest are other available states that the tachyon/particle may occupy. The tachyon/particle is most likely (Probably) going to end up in the available state that is closest (proximate) to it. **Tunneling** tells me that this probability is an exponentially decreasing function of distance.

The Properties

Size of a light cone (Available state?): Tachyons can interact when their light cones intersect. A tachyon, being faster than light, is always out in front of it's light cone. Curently, I define the size of a light cone as the distance it takes for the velocity to change by c. Assuming the tachyon is traveling in circles, it's velocity changes by c in [(loop radius) x (c) ÷ (Tachyon Speed)] meters, which is the length of it's light cone. I then have the light cone spread out laterally by c for the time it takes the tachyon to travel this length. So the light cones from the helix are very very tiny, while the light cones from the loops are very large.

The **Surprise** mechanism. Tachyons can knock each other out of existing available states kind of like billiard balls. They temporally displace each other, but their available states have them sticking around before reacting. If a particle is accelerating, it is changing available states. If the state the accelerating tachyon is entering already has a tachyon in it, then the tachyon already in that state is surprised by the entering tachyon. The faster the entering tachyon is moving, the closer it can get to the tachyon already there before the already there tachyon can respond. The already there tachyon is surprised by the entering tachyon. The greater the surprise, the closer the two tachyons get before the repulsion action kicks in and the greater the responding tachyon's velocity.

Energy: To conserve energy is to consider that every tachyon that is part of a real particle must contribute to the pressure of the cloud and it's pressure proportional to it's mass. Thus I will make all loop speeds the same then the helical circle will contain the velocity that describes the mass. (The tachyonic repulsion of the centers of mass might need a further condition of the force being proportional to the masses that are repelling. Again, the inverse r of the loop relation to energy described in the Shape appendix may preclude the need for this rule.)

A Tachyon Theory of Everything

Tachyon-Cloud Duality: The tachyon both generates, and is altered by it's available state. So I need the tachyon dynamics to match the cloud dynamics. I think there should be a Tachyon/Cloud duality in my mechanisms, and that this duality is probably an extension of the wave/particle duality in QM.

Distributions: I want the pressure and duality to result in some distributions in the cloud. For example, I need tachyons in the cloud that are not part of existing particles to be in ±im pairs to help produce the 2 tachyon photon. Also, The spinning multidimensional tachyon mass may force neighboring tachyons into the needed patterns for Quark and nuclear arrangements, the electronic structure around nuclei, and also QM stuff. In some cases there will be empty available states around an arrangement, in others, the available state will be much larger than the shape of the tachyon occupying it.

Tachyon Rest Mass: A link between the properties of the cloud and the tachyon is a variable tachyon rest mass. As the graph up front shows, there is no natural rest mass for tachyons. I am free to relate the rest mass to the pressure it is experiencing. But I must always be mindful of the constraint to keep all tachyon paths completing their loops. And I still assume them to all have the same rest mass under the same energy conditions. If I can relate the curvature of the path strictly to the velocity, then I can modulate the rest mass without altering the path.

The Physics

The PHYSICS; Describing the universe with the above properties.

Start with the idea that every real particle (excluding neutrinos which are available states with no tachyon in them.) is made up of at least one tachyon in at least one available state, and that these tachyons are traveling in loops that range in size from a galaxy to the whole universe.

Those tachyons that are not part of real particles can manifest as **virtual particles**.

We are in a big cloud of tachyons, many of which are from across the universe.

Inflation, and why it stopped: So right away we get the inflationary universe with a burst of infinite velocity tachyons from the zero energy void (They either come with a remnant of higher energy tachyons or they somehow loose or redistribute their energy to have higher apparent energies) that initially expands at tachyonic speeds until they can complete their loops (Initially the cloud may be too dense for them to come close enough to self interfere. This implies a maximum density that can exist to produce real particles. The Back Hole is only very dense for a small portion of the tachyon's path. Too big a Black Hole and physics could get odd.) to form real masses (wherever that real mass point on the loop may be.) that must be restrained to the real particle's speed. This could be an ongoing process wherever the void gets low enough in tachyon density, more just come into existence. Maybe it's just the case that nothing standing still (empty space) equals something going infinitely fast (Tachyons).

Dark Energy/ Cosmological Constant: The Temporal pressure from all the loops (The center of complex mass in the helix.) is Dark Energy pushing everything apart, but it is also pushing things inside it together, like nucleons and oppositely charged particles.

15

A Tachyon Theory of Everything

Shared Available States can explain the following;
    Strong Force: This type of shared available state can hold two baryon shaped tachyons at the same time. The Susskind Interpretation of the Veneziano model for the Strong Force is of a vibrating string. I'm thinking this string is a contiguous spacial extent of more than one available state that more than one tachyon may share simultaneously. And I'll bet rings of these available states form Helium nuclei. So Dark energy is shoving nuclei together in tiny bubbles of the tachyon cloud.
    Quantum Entanglement: is another type of set of shared available states that is not connected, so the two particles sharing them are free to drift apart.
    Also, **neutral particles** (except neutrinos) like **Neutrons** are shared available states containing the electron and proton shapes, and the **photon** is also sharing two oppositely massed tachyons to be neutral.

    Electricity and Magnetism: The Tachyon-Real mass interaction results in the idea that the flux lines coming out of a charge can be thought of as teeny-tiny time-space tornadoes. And like the Bernoulli Pressure reduction in a regular tornado, so the time space pressure (Dark energy) is reduced in a flux line. So charges move towards the region of greater flux density. And the helical action of any tachyon provides it with its' own electric field.
    The magnetic field is simply the relativistic electric field with added length contracted distortions on the charges' spheres. (See Appendix)
    My unifying force is a pressure. All physical entities have a finite size. Force is pressure times area. Thus the tachyons making up all my particles are of finite size. No point particles.

QM Probability Waves: The Available State is a probability wave. It is a place where you might interact with the tachyon's particle, but because the tachyon is going faster than light, you have no way of knowing where in that available state you will find the tachyon's particle.

Heisenberg Uncertainty Principle is between the properties of the available state (p, E, ...) and the tachyon's particle's reaction (x, t, q ...). By measuring the properties of the available state, you only know where it is possible to find the tachyon's particle. But since the tachyon is traveling faster than light, you have no way of knowing where in that state it will be. On the flip side, to measure the tachyon's particle's position or time is to interact with it for a finite time delay allowing the faster than light tachyons of the cloud to whip by creating an entirely new and unpredictable configuration in which the tachyon will carve out a new and unpredictable available state. Notice that it's easy to see how the wave function collapses throughout space instantaneously.

Single particle interference: The available state of the particle on the way to the double slit spreads out in front of the particle to guide it to a pattern on the screen. The particle really does go through one particular slit, but to measure which slit will involve a time delay, and that finite period of time will allow the cloud to change to a new pattern. In a way the available states are doing Feynman's sum over paths. Instead of having the particle travel over every possible path, I have available states map out all possible paths while the particle unpredictably follows just one.

Photon: First note that any vector can be split into perpendicular components that are smaller than it, and that one of the components can have an arbitrarily chosen magnitude as long as it's less than the vectors magnitude. Thus any tachyon can have a component whose magnitude is c. So let's orient an arbitrary tachyon such that it has an x-component of magnitude c, and the rest of it's velocity oriented along the y-axis and labeled U. Then the tachyon's speed, v, (No vector label and not bold means magnitude.) can be expressed as $v^2 = c^2 + U^2$ with the vectors;

$\vec{v} = (c, U)$, and $\vec{p} = m\vec{v} = \gamma m_o \vec{v}$, where $\gamma = \dfrac{1}{\sqrt{1-\beta^2}}$, and $\beta = \dfrac{v}{c}$,

then

$$\sqrt{1-\beta^2} = \sqrt{1 - \frac{(c^2 + U^2)}{c^2}} = \sqrt{-\frac{U^2}{c^2}} = \frac{iU}{c} \text{ , so } \gamma = \frac{c}{iU}$$

Concerned with magnitude only here, I'll ignore the i in the gamma term.

$$\vec{p} = m\vec{v} = \gamma m_o(c, U) = \left(\frac{c}{U}\right) m_o(c, U) = \left(\frac{m_o c^2}{U}, m_o c\right) = m_o c\left(\frac{c}{U}, 1\right)$$

So the component with the constant velocity (the c in the x-direction) has a variable momentum inversely proportional to U, while the component with the variable velocity has the constant momentum of $m_o c$. This to me screams photon with it's constant velocity but variable momentum and the Y directed **E** field changing with the X-directed momentum. I'm picturing the U component making a loop the size of a galaxy or maybe even the entire universe, and the loop is traveling at c.

I propose the following schematic (Below); The smaller horizontal component is c, and the longer vertical component is U. The solid line is +im and the dashed line is -im.

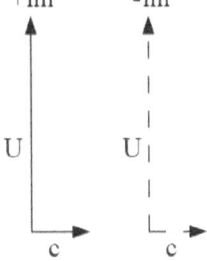

The Physics

I want the **photon** (to the right) to be made up of two oppositely massed tachyons to make it neutral. All of the dashed line arrows represents one -im tachyon on 2 legs of it's loop. Same for all of the solid line arrows representing one +im tachyon on two separate legs of it's loop. The left sets are half a wavelength from the right set. The top sets of arrows are simultaneous and superimposed with the bottom sets. With this overlap, the helical action of the vertical components reinforces while the helical action of the horizontal components cancel. Thus the angular momentum (**E**-Field) is strictly vertical and oscillating between pointing up and down.

Then for a particle like an **electron** (to the right), I will use just two legs of the same type (Dashed). The two legs are oppositely directed canceling each other out, but the single type (All legs are Dashed) means a charged mass.

Note: The only difference between the dashed arrows in the photon and the particle are that the c-component on the bottom arrow has switched direction, and the arrows to the right are superimposed while those in the photon are separated by half a wavelength. Note also how the particle to the right kinda looks like a photon bouncing back and fourth. Also note that you can't connect two opposing arrows tip to tail with circles, thus the loop may not be a circle. These models may be wrong, but I like the way the components add and cancel, so I'll stick with them for now.

A Tachyon Theory of Everything

Particles are bouncing Photons: Indeed, most of the odd relativistic properties of matter make more intuitive sense when modeled by photons bouncing back and fourth in a mass-less box. I say more intuitive because we have the natural experience of Doppler shifted sound waves. Although the formulas are a little different for the light waves, they still covey the intuitive notion that approaching an emitting source results in increased frequency/energy being received and receding from an emitting source results in decreased frequency/energy. Some of these properties include;

 Relativistic mass: Because of the square roots in the formulas, the increased Doppler shift is always greater than the decreased Doppler shift and the net is the relativistic mass increase (See appendix).

 Inertial mass: The sum of the photon's momentum shows it equal to the relativistic mass momentum.(See appendix),

 The limit of c for speed: If a mass is a photon bouncing back and fourth, then the center of mass' velocity is determined by what fraction of the time it's going forward at c versus the fraction of time going backwards at c, and since it must spend some time going backwards, it can never average to c.

 Mass-Energy equivalence Since photons are energy, modeling a mass with them is modeling a mass with energy.

 Force in Gravity is gotten from using the Gravitational Red Shift. The bouncing photon has less momentum bouncing off the top of the mass-less box compared to bouncing off the bottom with the difference being the net force of gravity on the box and it's contents of photons (See appendix).

The Physics

From the above it's kind of easy to visualize **pair production** in which a photon creates two oppositely charged particles. The interaction just flips one of the c components on each tachyon type so that both tachyons now look like the particle schematic and act like photons bouncing in their mass less boxes.

Annihilation requires involving a couple of tachyons from the cloud to get the required four tachyons to make two photons.

Again, what happens between the depicted branches in the schematics is still to be determined.

Gravity: A spacial density gradient of tachyon mass-energy in the tachyon cloud in the Void to produce the "stretchy - shrinky space" of Einstein's General Relativity. My model just explains that the "shrunk space" in towards the mass is caused by the tachyons from all the particles in the mass and that the increased density also increases the temporal pressure to increase the momentum/energy (by increasing the rest mass) of the tachyon making the particle of interest in the gravity field.

I can also use ppp to say the the increased density of tachyons towards the mass means the available states are squished smaller and therefore closer together. Thus the closest available state is also closer to the large mass, and favoring the particle "tunneling" towards the large mass. This helps equate gravitational and accelerational mass since both simply involve "spiraling" the tachyon path in a particular direction.

Gravitational Red Shift: The higher density of tachyons in towards the mass provide three things to a falling photon;

Smaller, closer available states shrinking the wavelength

More time tics increasing the frequency

Higher temporal pressure increasing the energy.

The wavelength consideration is another hint that the **Self Pressure** mentioned in the Shape appendix might be important.)

A Tachyon Theory of Everything

Notes on the two gradients; Just to be explicitly contrasting here. The **gravity gradient** is caused only by those tachyons belonging to particles in the near by mass. The energies of the tachyons contribute to the gradient, so the tachyons coming from protons contribute more than those coming from electrons, so the mass of the object matters. This gradient is sitting on a large (Deep?) field of cloud tachyons from everywhere else in the universe that has no gradient (or not much of one). The **electric gradient** is caused by ALL of the tachyons in the cloud that pass through the charge. Every tachyon gets the same ΔL independent of that tachyons energy, so only the number of tachyons matter and not their energies, then any two particles in the same region of space will have the same charge.

Also, if I allow the loops of protons to be about the size of galaxies, then most of the mass-energy-gravity flux ends quickly at the edges of galaxies creating a much large gradient than the inverse r squared law would calculate at the galactic edge. Higher gradient means larger gravitational force to explain **Dark Matter.**

From the Shape appendix, if the loop radius is inversely proportional to the energy, then the electron's tachyon's loops are a couple of thousand of times bigger than the proton's loops, and the microwave background radiation, providing two tachyons per photon, has loops many millions of times bigger. These large loops allow for the encompassing so many tachyons from so far away, that any tachyons contributed from a nearby planet or star is immeasurably trivial as a percentage of total tachyons making up the electric field gradient.

Superposition of gravity and electricity comes from tachyons traveling right through everything.

Proton: I think the proton has a complicated trajectory that may have 3 separate points of real mass interference for each of the **quarks** (See Shape appendix). And cloud tachyons may be involved to get the oppositely charged quark, or the oppositely massed tachyon may be the result of a relativistic view. The single +q charge of the proton versus the fractional charges of the quarks comes from a notion of "Time Sharing" the cloud tachyon collisions making the **E** field. The Real masses of the quarks are only there when the tachyon is passing through them, which can only happen once per loop and there are three loops, so not all the quarks are conveying a charge all of the time.

Also, the three quarks being the same tachyon explains their increase in energy as they are separated, as the tachyon making them up is stretched like a spring.

Neutrino is an electron shaped available state with no tachyon in it.

Neutrinos are the only neutral particle with no tachyons in them. All other neutral particles have both types of tachyons canceling each other out. The filling of available states with two tachyons results in lifetimes for neutral particles that are generally shorter than that of charged particles due to the tachyons not being connected, and thus ultimately drifting apart.

Anti Particle: The particle made of the oppositely signed ($\pm im$) tachyon. Note that I rearrange the whole notion of antiparticle. My proton and electron are made of oppositely signed tachyons, but they are not anti w.r.t. each other. Their tachyons have very different velocities and are way outside each other's light cones so they do not interact. (Thus an S orbital electron can pass right through a nucleus.) When two identical shapes of opposite tachyon sign interact, their angular momenta "unwind" leaving photon shaped tachyons.

Virtual Particles: Those tachyons in the cloud that are currently not in self-interfering shapes are available to be turned into real masses. Although the complex part of the trajectory cannot be interacted with, the "ends" of the loops that would self interfere can be manipulated to alter the shape of the tachyon to make real particles.

All **unstable particles** are sums of the stable ones. I have a notion that the stable patterns of electron, proton, and photon (and their anti particles) form a basis set of "vectors" and all of the other particles are linear combinations of the basis set.

Bosons are fast particles like photons. Going at or near the speed of light puts them on the edge of everything's light cones thus reducing their temporal pressure by shrinking their light cones and allowing more to fit in a box.

Conservation of Lepton and Baryon number is just the conservation of angular momentum of the tachyons of the cloud.

For example, in **Pair Production**, the creation of an electron and a positron, two tachyons of opposing handedness in their angular momenta each carry the opposite angular momenta of the other.

For anoyher example a **Neutron decays** to a Proton which is fundamentally the same shape, by kicking out two oppositely spinning electron shaped available states, but one is filled with a -im tachyon (the electron) while the other spins off the opposing angular momenta in the cloud as the electron's tachyon, but lacking the extra tachyon, remains empty (the **antineutrino**).

The **Mesons** are double states where each one contains available states with opposing angular momenta precluding the need for extra particles to balance momenta.

Pauli Exclusion Principle: Is really a force when looking at it's role in neutron stars. This is produced by the temporal force between the tachyons themselves in their helical circles. The tiny helical circles limits the size of the light cones and thus the distance the force is felt over. The Exclusion force only acts on similar particles of similar energy where their tachyon's speeds are close enough to be in each other's light cones and repel.

The Physics

An **S-Orbital** electron's tachyon velocity is so different from that of the protons and neutrons in the nucleus, that they are outside each other's light cones and don't even "see" each other, so the electron flies right through the nucleus as if it's not there.

Two **electrons flipped w.r.t. each other** will have their respective tachyons looping and spinning in opposite directions allowing them to fit together without having their tachyons share a velocity to repel each other.

Weak Force is not really a force: For this one I revert to it's original interpretation as the Dirac Interaction. In my theory the two shapes of the incoming particles are exchanging angular and linear momenta to change into the two outgoing particles.

Energy of decayed products. I learned in school that when all the energy is summed before and after a decay, the difference is available for kinetic energy, but nobody ever explained how the particles or even the space between them "know" what energy is available for motion. The temporal pressure between the tachyons making up the decayed products is what gives the decaying particles their kinetic energy. Most particle's tachyons are in available states in the cloud and feel equal pressure from all sides, but the products of decay are initially formed outside of available states (or both are in the same unsharable state.) and so feel the net force between themselves while the cloud pressure is uniform all around.

Magnetic Moment: The Tachyon-Real mass interaction mentioned earlier must have a preference for interacting with those tachyons moving and rotating with the charge. Thus a moving charge creates a moving electric field and a spinning charge creates a spinning electric field. Also, note that I have assigned mass to the Available State while saying charge is a tachyon property. I picture all tachyons as being the same size, and the self interfered real mass is a sphere that sits completely inside the available state.

Thus the charge and the mass are not the same physical entity, So there is a degree of freedom in calculating the magnetic moment as a function of it's angular momentum. And with tachyons the model is not constrained to have its spinning surface move at less than c.

Precession is one of those interesting shapes that can be swept out by the tachyon. Then the local available state looks like a cone that is being swept out by a vector along it's straight edge. This prevents orientation along the tachyon's vector (and thus along a specific orientation to the particle.). and any attempt to measure it will again invoke time delay that will cause a new cloud configuration to follow.

Free particle-photon interaction: **Compton Scattering**: When the charge is accelerated by the photon's electric field, the charge's tachyon is lifted out of its current available state and thus shoved into a new one. The speed at which the charge's tachyon is accelerated determines how "surprised" all the tachyons in the interaction are. What I mean by that is the speed at which a tachyon is sent into an occupied available state determines just how close the two tachyons can get before the reacting one accelerates away. The bigger the "surprise", the closer the tachyons ultimately get so the greater the temporal pressure and thus the greater the acceleration and the higher energy the reaction.

So first the incoming photon removes a charge from it's available state thus at the very local level the cloud pressure has been reduced by the opening up of an available state. This lower pressure lowers the energy of the incoming photon out of existence. But now the electron has just been shoved into a new available state that is probably occupied by cloud tachyons and thus they are kicked out and pressured into being a photon.

Cherenkov Radiation: By going faster than light in a particular medium, the charge is entering new previously occupied available states with each new position, and kicking out the tachyons in those positions to be photons. The draining of available states along the path lowers the temporal pressure and thus the energy of the charge.

Tachyons have no charge being complex tachyonic particles flying right trough everything, only the slow moving Real points of self interference have the charge. So no Cherenkov paradox.

Pair Production: As mentioned earlier when I drew the schematic for the photon, just by reversing one -im c-component and one +im c-component you get the schematic of two particles whose tachyons are now bouncing back and fourth like a photon in a mass-less box.

Annihilation: The electron and positron being made of oppositely handed spinning and orbiting tachyons can unwind or cancel some of their angular momenta. The tachyons having been changed to new available states that are probably already filled, those tachyons being displaced create the new photons. Note that I have 2 tachyons coming in, one from each particle, but 4 tachyons going out with 2 per photon. This requires a distribution in the cloud of two opposing tachyons being displaced by each incoming particle's tachyon.

Momentum: A particle's available state is spread out around it and has empty parts that the particle can occupy. In the particle's rest frame those states are equally spaced around it, and the particle randomly jumps around (On a very microscopic level). In a moving frame those states are Doppler shifted and those out front are closer while those to the rear are farther apart, thus with PPP the tachyon making the particle has a greater probability of ending up in a forward state that a rear state. The greater this Doppler shift, the greater the probability for the particle to move forward, thus the higher momentum. So I have momentum being related to the probability for the particle to appear in the next available state along it's trajectory.

A Tachyon Theory of Everything

I can use this probability argument to help **conserve momentum**. For two particles at the moment of collision, in a uniform cloud, and in the center of momentum frame, the two closest available states will probably be on opposite sides of a sphere centered on the collision. Similarly, for a photon particle interaction, the excess tachyon that will participate in the photon is probably opposite the empty available state for the particle.

Tunneling: A particle doesn't tunnel through barriers, it disappears into it's constituent tachyon, travels around the universe, then reappears on the other side of the barrier. But then again, in my model the particle is disappearing and reappearing or tunneling all along it's trajectory weather it's tunneling through an energy barrier or not.

Black Holes have had the temporal pressure from the helix action (Exclusion Principle pressure) overcome and all of the real self interfered masses are in the same spot, which is fine since tachyons can all fly right through each other. The tachyons of these masses are still flying about the universe, and tachyons from around the universe are still flying right through it, so a black hole still has the gravitational and electric fields.

The **Event Horizon** is where the tachyon density is such that subsequent tachyon tics occur before the particle can move (The next tachyon is entering the particle before a previous tachyon has left.), so infinite time will pass before the particle advances further. With this I imply that in normal space the tachyons of the cloud hit the particle one at a time with tiny gaps between collisions allowing the particle to move.

Big Bang: Explosion of infinite velocity tachyons from the empty, zero energy void. Looking at the curve on the front page for tachyons, there may be no physical difference between infinite velocity tachyons, and the asymptotic limit of zero energy space. If this is an ongoing process, then the new bursts could happen with the tachyons at any velocity, thus there may be pockets of the universe with velocities out of sync with their surroundings.

More thoughts

Atom/Photon Interaction: For absorption, the photon's **E**-field interacts with the electron to both move the charge to an empty available state, and provide a time delay to allow for a new configuration in the field thus allowing the tachyons in the photon to "occupy" the electron's old available state. Not sure of the nature of this "occupation", just need tachyons present to allow; For **emission,** the electron, with it's tachyon, is dropping into an "empty" state. In doing so it "surprises" the tachyons in that state and they are pushed out with the resulting pressure of the push determining the emitted photon's energy. The greater the energy drop of the electron, the faster it's speed when entering the occupied state, the greater the "surprise" ,and thus, the greater the emitted photon's energy. Also, by traveling away at c, the emitted photon is keeping it's tachyons outside the light cone of the tachyon from the charge that "surprised" them in the emission process, which is in keeping with axiom 1 with tachyons being tachyonic w.r.t. each other.

Proton: The shape of the tachyon path has 3 points of self interference to match the 3 quarks. The points of self interference are not complete. The complex masses of the three points have not fully canceled into scalars and the resulting masses are still extended in such a way that they are swept into each other such that this final sweeping into each other is what produces the fully interfered scalar masses. In this swirl of a self interfering tachyon, I allow an oppositely masses tachyon from the cloud to partially take part to create the negative quark. But it could also be the case that the negative is coming from the relativistic view of one of the points looking like a negative massed tachyon. The quark's fractional charge is a result of "time sharing" the single fully real massed tachyon's single charge among the three points. It could also be that the 1/3 tachyons are swept into each other to create 2/3 charged masses, then the single tachyon from the cloud contributed it's whole negative charge on one point to make it -1/3.

The sweeping motion may be part of the contiguous available state of the string mentioned earlier.

That all 3 quarks are made of the same tachyon, trying to pull them apart amounts to trying to stretch a spring, the farther apart the quarks, the further stretched is it's spring and the greater the force trying to restore things to the way they were. Also as the ends spread apart, they push aside the cloud creating a larger bubble of an available state thus more force from the three points is needed to maintain the larger bubble.

I can imagine a couple of mechanisms for the jets of particles streaming out from attempted quark scatterings 1- Stretching the tachyon's ends apart leaves truly empty void in between for tachyons to be brought fourth from the zero energy void. Or 2- that virtual particle-tachyons can fall into the bigger available state to become real particles.

Momentum: For the cloud/pressure point of view, the tachyon is in a type of energy well when it's in it's available state. The walls of the well/available state have a slope to them. That is the tachyons of the cloud exert a pressure on the tachyon and this pressure has a spatial ramp to it. So when the tachyon undergoes an interaction, the time delay of the interaction causes the particle to freeze while the cloud and the available state keep moving until the time delay is complete and the particle can respond. The combination of the speed, time delay, and "energy well ramp slope" work to press the particle further along it's trajectory. Thus the gradient of the available state's wall is related to momentum.

The Physics

Infinite self energy of a charge is taken care of with the finite size of the self reinforced tachyon generating the field.

Free Space Fields are simply a region of space with a net angular momentum in the tachyon field.

Fields in a photon: The **E** is the reinforced angular momentum directed perpendicular to the velocity. The **B** field is the relativistic **E** field when seen flying by at c.

The imbalance of matter and anti-matter is an imbalance of angular momentum, but it might not be an imbalance if the extra angular momentum can be taken up by some other imbalance of angular momentum like say an observable difference in the relative amounts of left and right circular polarized light, or even the universe itself having a spin to it.

Further Ramblings

Another possible manifestation of the 3-D nature of the complex mass is that each mass-dimension may have an orthogonal reaction w.r.t. the other dimensions. 1) Front and Back are already scalar due to the back flying into the front to cancel out that extension. 2) Inside and outside the loop masses react "temporally" by speeding up or slowing down along their trajectories. 3) Left and Right react spatially to get a sweeping action on the tachyon's path. So one axis reacts temporally, another reacts spatially and the third is scalar and orthogonal to the other two dimensions.

Another way to get a torus for the basic particle shape is to have the tachyon making it up undergo a rotation (Delta L) when passing through it's own point of self interference. Then the torus is swept out one loop at a time. This might fit nice with the whole charge / field model.

Another loop shape to consider is that of the cardioid. (See Shape appendix)

A potentially useful distribution of tachyons around an atom is to have an arrangement where the density of tachyons inside the atom is lower while just outside it's higher. Then I can terminate photons inside the atom and rationalize lower masses for bonded particles.

If tachyon loops are not the size of the universe, then when traveling between large chunks of mass, like between galaxies, where that tachyon density may go down considerably, we may find ourselves passing through the void much faster than expected, and physics might get weird.

It may be the case that the more energetic a particle, the smaller it's loop size is. This may help with the exclusion principle, allowing a protons exclusion to be smaller than an electron's. Also, applying the same concept to the loops, making the proton loops the size of a galaxy while the electrons are several thousand times bigger (and the loops from microwave radiation tachyons being larger still) might explain dark matter while still allowing for spherically symmetric **E** fields around charges near large masses.

Temporal Pressure and Mass-Energy

From the curve on the first page, the energy of a tachyon is related to it's speed. I'm relating energy to pressure so pressure should relate to speed, which fits with slower tachyons having more energy since they are around longer to exert their contribution to the pressure.

The spread of velocities related to the temporal extent/angular velocity of a tachyon looks like a contributor also since a larger spread of velocities will in turn displace more velocities in the cloud. Perhaps I should include this in the size of an available state.

Using $V = R\omega$, and the fact that for a single tachyon all parts of it are moving at the same ω, define;

R_i is the inside radius of the loop, V_i is the velocity at that radius
R_o is the outside radius and, V_o is the velocity at that radius
R_t is the average loop radius, V_t is the velocity at that radius
and is the same for all tachyons.

Let $\Delta = R_o - R_i$ to be the size of the tachyon which is the same for all tachyons,

So the velocity difference or spread is

$$V_o - V_i = (R_o)\omega - (R_i)\omega = (R_o - R_i)\omega = \Delta\omega = \Delta(V_t)/(R_t)$$

So, since both V_t and Δ are the same for all tachyons, the spread of velocities is proportional to $1/(R_t)$

To conserve energy, I think I need all loop velocities to be the same (or within c of each other.) and the temporal pressure between them to be proportional to the mass-energy making them up. But the loops can still be different sizes. Doppler shifted photons demonstrate that the energy of a particle must be independent of the loop radius since that does not change perpendicular to the velocity, but I can claim that the loop size in the rest frame of the particle's generation is dependent on the energy of the particle, which I do in the Shape appendix.

Just to be "outside the box" another possible contributor to pressure would be the notion that for tachyons, the velocity is their position coordinate making acceleration, in this case angular acceleration, the velocity and the tachyon can oscillate between angular velocities as it "bounces" off the other tachyons in the cloud. Basically I'm picturing the tachyon bouncing off the walls of it's available state, where the frequency and energy of the collisions contribute to the pressure. Must prevent the potential spiraling this mechanism can create. Since $r = v/\omega$, the tachyon's velocity must be allowed to vary for this to be a mechanism (vary within c?) Also use average velocity when looking at this?(But I ignore this for now.)

The temporal pressure between two tachyons each in their own independently oriented curved orbits, but momentarily have the same velocity looks something like a bell curve going from zero when they are greater than c apart, then climbs in temporal force to a peak related to just how parallel they got, and how far apart they are, the mass/energy of each loop and their spatial distance from each other, then drops back down to zero as both tachyons continue around their loops and become greater than c apart in velocity.

Further Ramblings

Some odds and ends;

Not sure if tachyons even see anything as they travel through the void except the occasional real mass whose electric field they take part in.

With available states and tachyons popping in and out of them, one model I think about is that space is kind of like a semiconductor with the cloud making the semiconductor, the particles are the charges, and available states are the holes.

-it? If tachyons are traveling in negative complex time, I might get two more for a total of four types of tachyon with +/-im and +/-it, but this complicates things more than simplifies them right now, so I'll ignore -it. Also, I started by considering the masses both because that's what the equation spits out and I was hoping to get a negative linear momentum to explain the attractive electric force. Having rejected negative momentum, I now think my two tachyon types might be the ±it while ignoring the -im. Or maybe they are +im, +it, and -im, -it, and I ignore the crossed masses and times. No matter, for now I need two opposing tachyons that I have labeled +im and -im.

The relativistic connection between mass and time (velocity) might mean that the same phase condition that creates a real scalar mass also creates a real scalar time.

Finally, A Little Defensiveness, besides the fact that there's little math, and I avoided the complex i.

Some would say this is all unproven assumptions and easily dismissed! And taken individually I would agree (Which is why I haven't published anything till now.), but each assumption has the variables going in the right direction. And together they explain everything. And the model is rich enough to provide the degrees of freedom needed to work out the details and solid enough to reject ideas that don't fit.

A Tachyon Theory of Everything

A LOT of structure for one little particle. I have bestowed tachyons with six sides. Not only that, but different sides have different masses and time frames. I have them acting at a distance when temporally displacing each other (but I do have them touching in time.). I have a set of collision rules for tachyon-tachyon, tachyon-real mass, and even colliding centers of mass. But I contend that this is hardly more structure than that proposed in String Theory with their strings (of what?) under tons of tension (What's pulling on them? And why don't they collapse to zero, so what's pushing against the tension? Actually I think this tension is the Dark Energy cloud pressure acting on the outside of the available state.) Then there are all of those dimensions of string theory. I'll stick with my 3-D timeless void. I know I have multiple complex time and mass dimensions, but I've got interpretations of these. Nobody has such an interpretation of extra spatial dimensions (Curled up? That to me is crazier than tachyons.). In freshman physics we learn that the ideal mathematical structure of a projectile is an up-side-down parabola that stretches to and from $\pm\infty$, but then I look up from my paper to view the reality of the ground and back on the page, I lop off the ends of the parabola to match reality. Same deal with String Theory and its 11 dimensions (or 24, or whatever their up to now.). Sure the original String answered some questions that I've actually incorporated into my theory, but to run with the idea all the way into 11 dimensions is where I look up from the page and see only 3 dimensions so I lop off the extra dimensions.

Note how I can add up (3 space dimensions) + (1 Real time dimension) + (3 Complex time dimensions) + (1 Real mass dimension) + (3 Complex mass dimensions) to get a total of 11 dimensions. Maybe any 11 dimensional model that can support a mechanism for waves will look like a promising model for the universe.

Further Ramblings

Only being able to interact with the Real "end points" of the tachyon's trajectories precludes the ability to do faster than light communications with them. So no time paradoxes.

A little aside question regarding the direct application of Relativity. The Inflationary model was rationalized on the notion that, since the parts of the universe that were flying apart from each other at speeds greater than c never observed each other, they weren't restricted by relativity. Does this mean that I can't apply relativity to tachyons, since I will never observe them?

So, everything seems to fit, and between the Void with Gravity, Available States with Uncertainty and the Strong Force, and helical action for the E&B fields and Exclusion, I think I've at least notionally tied it all together with some pretty neat ideas. Besides, I could keep this to myself for the next several years while I do something original like figure the masses of the leptons, but I think I should share what I have and I'd like to try to monetize my effort, so here it is and I hope I described a comprehensible and clear enough idea to inspire and amuse you the way it has me.

A Tachyon Theory of Everything

The Electric Force comes from the Tachyon-Real Mass Interaction, and the notion that the Bernoulli pressure reduction in a wind can apply to the Tachyon Cloud's Temporal Pressure (Dark Energy) being reduced in a region of a similarly recognizable "wind".

   The typical tachyon's trajectory is shaped like a long slinky bent in a loop, but the loop is so large that it is locally a straight line. Separate the momentum and kinetic energy into a rotational component for the action around the tiny helical spiral, and a linear component for the spiral's center of mass traveling along the loop. So this locally straight line is like a teeny tiny time-space tornado, the wind mentioned above.

   When a tachyon travels through a real, self interfered tachyon (a charge) it exchanges a (universally constant) little bit of its rotational kinetic energy for its linear kinetic energy keeping it's total energy constant. So the tornado is either increased or decreased in it's ferocity.

   To keep all charges the same, I keep all tachyons the same size. The available state they occupy and that conveys it's energy and momentum can be much larger. By making the charge the self interfered tachyon, then in a cloud of tachyons of a relatively uniform density, all charge's will have the same number of tachyons going through them, and by making "the change in" (or Δ) rotational kinetic energy universally constant, then all charges will have the same number of tachyons coming out from them with the same Δ rotational kinetic energy. So coming out of a charge are a bunch of right handed tornadoes (+charge), or left handed tornadoes (-charge).

   I'm focusing on energy above since pressure is energy density, and it is ultimately temporal pressure pushing the charges around. Since energy is not a vector, and angular momentum will make the next discussion much easier, I am simply going to associate angular momentum, **L**, with rotational kinetic energy. When the rotational kinetic energy from a +**L** situated tachyon is canceled by another tachyon's -**L**, then there is also no rotational kinetic energy to split out from the total cloud pressure's contribution to the energy density, and therefore no Bernoulli type pressure reduction from what are now canceling teeny tiny time-space tornadoes. Basically I reinterpret flux lines as net angular momentum in space-time.

+im tachyons have +**L** for their helical angular momentum. **L** points in the same direction as it's velocity along the loop or locally linear direction.

-im tachyons have -**L** for their helical angular momentum. **L** points in the opposite direction as it's velocity along the loop.

All tachyons traveling through a +im self interfered tachyon (+charge), get a + Δ**L**. That is;
+im tachyons get + Δ**L** and increase the magnitude of their **L**
-im tachyons get that same + Δ**L** but since there's is negative, the magnitude of their **L** decreases.

Flipping it all around for -im self interfered tachyons (- charges)

All tachyons traveling through a -im self interfered tachyon (- charge) get a − Δ**L**. That is;
+im tachyons get - Δ**L** and thus decrease the magnitude of their **L**.
-im tachyons get that same - Δ**L** so the magnitude of their **L** increases.

In the next two pages I walk through the interactions. First looking at two opposing charges interacting with each other, then how each charge responds to an external field.

+/-im are next to the velocity vectors. L's indicate angular momentum.

For tachyons coming from the left.

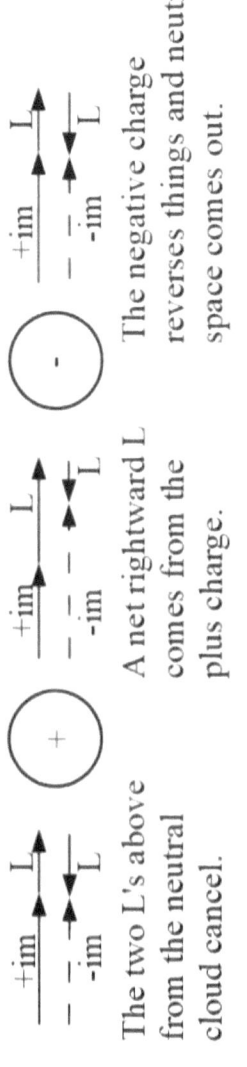

The two L's above from the neutral cloud cancel.

A net rightward L comes from the plus charge.

Plus charge reverses things back to neutral.

The negative charge reverses things and neutral space comes out.

For tachyons coming from the right.

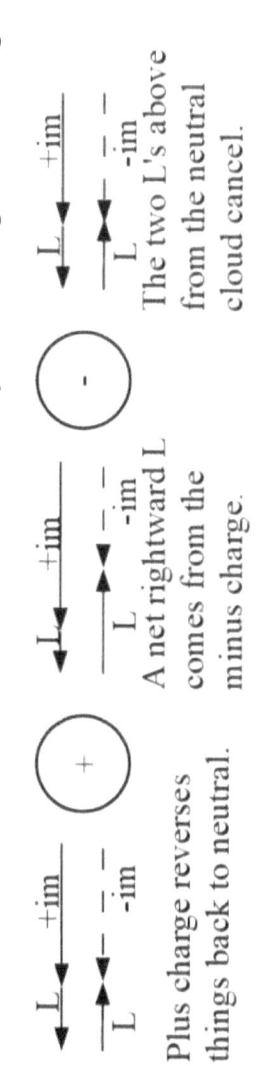

A net rightward L comes from the minus charge.

The two L's above from the neutral cloud cancel.

Both cases have a net rightward L between the charges thus a net "wind" to lower the cloud pressure (Dark Energy) and get pushed together.

A uniform external E field or **L in the tachyon cloud** pointing to the right

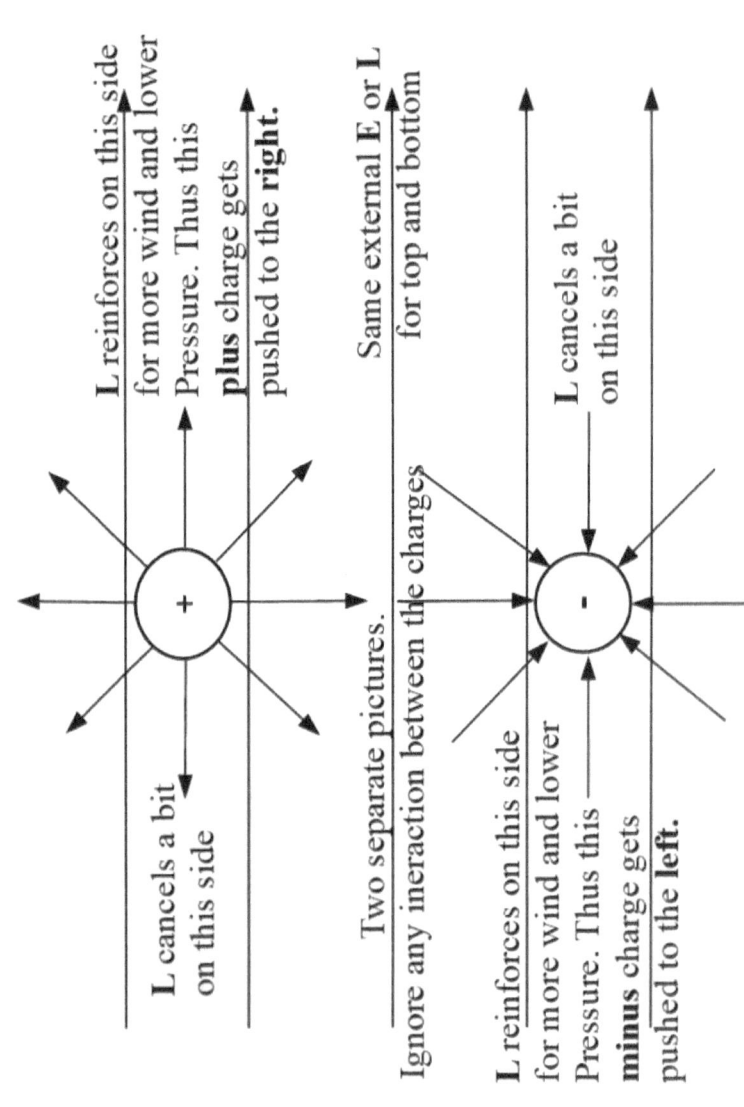

L reinforces on this side
for more wind and lower
Pressure. Thus this
plus charge gets
pushed to the **right.**

Same external E or L
for top and bottom

L cancels a bit
on this side

L cancels a bit
on this side

Two separate pictures.
Ignore any ineraction between the charges

L reinforces on this side
for more wind and lower
Pressure. Thus this
minus charge gets
pushed to the **left.**

41

A Tachyon Theory of Everything

A few more points, and the rules they imply;

The size of the self reinforced tachyon is too small to have been measure at this point (Plank length?).

The requirement to keep the energy the same through the interaction comes from the notion that the tachyons passing through the charge may come from stable particles somewhere else in the universe. Therefore any interaction must preserve that distant particle's properties. So on top of the total energy being held constant, there is also the requirement to preserve the "phase" of that tachyon's self interfered point. That is the point of self interference of any real mass is dependent on the relative orientations of the two constituent complex masses. So as the tachyon travels around the universe, it must preserve both it's total energy and it's phase or the orientation it will have upon arrival back at it's self interfered point. **Thus as the L of the helix changes, so must the radius of the helix to keep it's orientation at it's self interfered tachyon constant.**

That I have the loop velocity changing and a requirement that all tachyons have the same loop velocity or at least can be in each other's light cones means that the **delta velocities involved in the interaction must be less than the speed of light.**

To have **E** fields traveling and spinning with a charge means that there has to be a preference to the given ΔL to those tachyons traveling and spinning with it. This can be an all or none mechanism or some smooth function that averages out to the observed field.

Anyway, the charge in an external field model shown above is all that's needed to verify the electric field since any physical situation will look like that on a local enough level. And my mechanism makes for an inverse r squared magnitude to the force. And that tachyons go right through everything gives the superposition principle.

Appendix Electric Force

To get a little more rigorous, I want to get a pressure difference by looking at the difference in energy densities between the two sides of a charge in an external field. The two sides being that facing towards the field and that facing away. In the following picture the two sides are the top and the bottom;

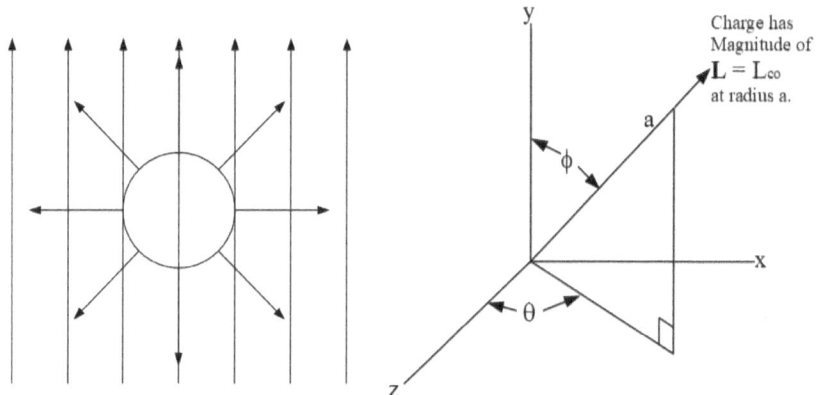

Charge has Magnitude of $L = L_{co}$ at radius a.

Define the vertically oriented vectors as the Background angular momentum, $L_b = (0, L_{bo}, 0)$ per meter squared. Define the vectors radiating out from the sphere of radius "a" as Charge angular momentum per square meter with y-axis vertical, azimuth angle ϕ, and rotation angle θ, $L_c = L_{co} (\sin\phi\cos\theta, \cos\phi, \sin\phi\sin\theta)$ angular momentum per meter squared.

On the top of the sphere, the Charge's angular momentum adds to the Background while on the bottom it subtracts. Thus the top has more angular momentum (or energy density) that is NOT contributing the the dark energy pressure compared to the bottom thus upward pressure pushes the charge up.

To combine the background's and charge's vector densities requires them to be expressed as densities w.r.t. the same areas. Then the vector combinations will all be per the same meters squared.

The special case of turning the charge's vector density on the surface of a sphere of radius r into a planar density parallel, in this case, to the background's field, is made easy with the following observation;

The spherical area element swept out by the radius vector above is $2\pi r \sin\phi$ for the length and $rd\phi$ for it's width while the planar element in the (x, z)-plane has the same length but a width that is only $rdr\cos\phi$. So, while the y component of the charge's field falls off as $L_{co}\cos\phi$, the area it is divided by is equally shrunk by the same $\cos\phi$ term cancelling the $\cos\phi$ and leaving the originally specified L_{co} for the planar expression. So, for the purpose of the dot product between the charge and background fields, the surviving y-component terms can simply be expressed as $\mathbf{L_{co}} \cdot \mathbf{L_{bo}} = L_{co}L_{bo}$ angular momentum squared per meter squared.

To calculate the energy on the surface I will use the usual rotational kinetic energy from undergraduate physics with $L = I\omega$ where the I is the average moment of inertia of the cloud tachyons, so;

$$E = \frac{1}{2}I\omega^2 = \frac{I^2\omega^2}{2I} = \frac{L^2}{2I} = \frac{\vec{L} \cdot \vec{L}}{2I}$$

Plugging in the charge and background values mentioned above;

$$E_{top} = \frac{(\vec{L_b} + \vec{L_c}) \cdot (\vec{L_b} + \vec{L_c})}{2I} = \frac{(\vec{L_b} \cdot \vec{L_b} + 2\vec{L_b} \cdot \vec{L_c} + \vec{L_c} \cdot \vec{L_c})}{2I}$$

and,

$$E_{bottom} = \frac{(\vec{L_b} - \vec{L_c}) \cdot (\vec{L_b} - \vec{L_c})}{2I} = \frac{(\vec{L_b} \cdot \vec{L_b} - 2\vec{L_b} \cdot \vec{L_c} + \vec{L_c} \cdot \vec{L_c})}{2I}$$

so the energy difference is:

Appendix Electric Force

$$E_{top} - E_{bvottom} = \frac{4\vec{L_b} \cdot \vec{L_c}}{2I} = \frac{2\vec{L_b} \cdot \vec{L_c}}{I} =$$

$$\left(\frac{2}{I}\right)(0, L_{bo}, 0) \cdot L_{co}(\sin\varphi\cos\theta,\ \cos\varphi,\ \sin\varphi\sin\theta)$$

$$E_{top} - E_{bvottom} = \left(\frac{2}{I}\right)(0 + L_{bo}L_{co}\cos\varphi + 0) = \left(\frac{2}{I}\right)L_{bo}L_{co}\cos\varphi \equiv dE_y$$

This is the energy difference in the y-direction per meter squared. By symmetry the energy differences in the x, and z- directions are zero,

so $dE_x = dE_z = 0$.

The force is the negative of the energy gradient, but I throw in another negative since this energy is subtracted from the uniform energy density of the dark energy cloud. With dy for the sphere being $2a\cos\phi$;

$$\frac{d\vec{F}}{dA} = \left(\frac{dE_x}{dx},\ \frac{dE_y}{dy},\ \frac{dE_z}{dz}\right) = \left|0,\ \frac{\frac{2}{I}L_{bo}L_{co}\cos\varphi}{2a\cos\varphi},\ 0\right| = \left(0,\ \frac{L_{bo}L_{co}}{Ia},\ 0\right)$$

The radius, a, and the moment of inertia, I are both constants. Integrating the element of force per meter squared over the area of the charge will just multiply by the maximum cross sectional area parallel to the background field of the charge which is, πa^2, putting the "a" on top. This uniform force acts like a uniform pressure across the cross sectional area of the charge.

$$\vec{F} = \left(0,\ \frac{\pi a L_{bo}L_{co}}{I},\ 0\right)$$

Notice the constant radius of all tachyons is required to

give all charges both the same field (allowing the same number of cloud tachyons to pass through), and the same force in an external field (same cross sectional area.).

Remember the units of the $L_{bo}L_{co}$ term is angular momentum squared per meter squared. So each L has units of angular momentum per meter. Assuming the magnitude of the angular momentum per meter is proportional to the magnitude of the electric field, let L = (Const)E. with the background being produced by charge, q_1, a distance r away (ignoring the distance, a, adds to r.), and the charge's, L_{co} produced by a charge, q_2, a distance "a" away. So plugging in the usual formula for the electric field;

$$\vec{F} = \left(0, \frac{\pi a}{I}(\text{const})\frac{kq_1}{r^2}(\text{const})\frac{kq_2}{a^2}, 0 \right) = \left(0, \left(\frac{\text{combined}}{\text{constants}} \right)\frac{q_1 q_2}{r^2}, 0 \right)$$

This all seems like a simple picture for a stationary charge in a stationary electric field.

But what about the magnetic field?

The magnetic Force . I will walk through some problems to work out the magnetic force.

I will consider the relativistic distortion of the charge's sphere into an ellipsoid as seen by the observer. And due to the charges velocity, I know it's observed ellipse is experiencing a relativistically concentrated and aberrated electric or angular momentum field. The field's concentration and aberration comes from considering what happens to the source of a uniform electric field as it moves relative to the charge. Thus, remembering from undergraduate physics that a uniform E-field comes from an infinite plane of source charge, I will be looking at the velocity of the infinite plane of source charges, **P**, and the perpendicular velocity of the flux lines, **fo**, along with a useful vector gotten by adding the charge's velocity parallel to the flux to **fo** that I will call flux velocity, **f**.

For each of following examples, the Traditional answers are in the appendix. Also, The prime mark, ' or double prime mark ",may be either on the variable or the subscript to indicate s'-frame or s"-frame. With c = Light, $v/c = \beta$ and $\gamma = 1/\sqrt{1-\beta^2}$,along with length contraction, where for a given length, $\mathbf{L}$, $L_{x'} = L_x/\gamma$, $L_{y'} = L_y$ and $L_{z'} = L_z$, in the next three cases I will be using the following equations from Resnick;

Table 2 – 2 The Relativistic Velocity Transformation Equations

$$U'_x = \frac{U_x - v}{1 - \frac{U_x v}{c^2}} \quad , \quad U'_y = \frac{U_y\sqrt{1 - \frac{v^2}{c^2}}}{1 - \frac{U_x v}{c^2}} \quad , \quad U'_z = \frac{U_z\sqrt{1 - \frac{v^2}{c^2}}}{1 - \frac{U_x v}{c^2}}$$

$$U_x = \frac{U'_x + v}{1 + \frac{U'_x v}{c^2}} \quad , \quad U_y = \frac{U'_y\sqrt{1 - \frac{v^2}{c^2}}}{1 + \frac{U'_x v}{c^2}} \quad , \quad U_z = \frac{U'_z\sqrt{1 - \frac{v^2}{c^2}}}{1 + \frac{U'_x v}{c^2}}$$

Equations 3-32 a & b, The Force Transformation Equations

$$F_x = \frac{F'_x + \left(\frac{v}{c^2}\right)\vec{U'}\cdot\vec{F'}}{1 + \frac{U'_x v}{c^2}} \quad, F_y = \frac{F'_y\sqrt{1 - \frac{v^2}{c^2}}}{1 + \frac{U'_x v}{c^2}} \quad, F_z = \frac{F'_z\sqrt{1 - \frac{v^2}{c^2}}}{1 + \frac{U'_x v}{c^2}}$$

$$F'_x = \frac{F_x - \left(\frac{v}{c^2}\right)\vec{U}\cdot\vec{F}}{1 - \frac{U_x v}{c^2}} \quad, F'_y = \frac{F_y\sqrt{1 - \frac{v^2}{c^2}}}{1 - \frac{U_x v}{c^2}} \quad, F'_z = \frac{F_z\sqrt{1 - \frac{v^2}{c^2}}}{1 - \frac{U_x v}{c^2}}$$

And equations 4-5 and 4-7 for the **E** and **B** transformations;

$$E'_x = E_x \quad, E'_y = \gamma(E_y - vB_z) \quad, E'_z = \gamma(E_z + vB_y)$$

$$E_x = E'_x \quad, E_y = \gamma(E'_y + vB'_z) \quad, E_z = \gamma(E'_z - vB'_y)$$

$$B'_x = B_x \quad, B'_y = \gamma\left(B_y + \frac{v}{c^2}E_z\right) , B'_z = \gamma\left(B_z - \frac{v}{c^2}E_y\right)$$

$$B_x = B'_x \quad, B_y = \gamma\left(B'_y - \frac{v}{c^2}E'_z\right) , B_z = \gamma\left(B'_z + \frac{v}{c^2}E'_y\right)$$

As a general rule for all my cases;
The s-frame is the ground frame and has a static vertically oriented **E** field, and no **B** field. The coordinate system is right handed with the x-axis pointing to the right, the y-axis pointing up, and the z-axis pointing straight out of the page.

Transform to the s'-frame moving to the right at speed v.
E = (0, E$_o$, 0), **B** = (0, 0, 0) and so the force on the charge is
F = (0, qE$_o$, 0)

Then for the s'-frame **E'** = (0, γE$_o$, 0) and
B' = (0, 0, γ(B$_z$ - (v/c^2)E$_y$) = (0, 0, - (v/c^2)γE$_o$)

For equations above, the velocity being transformed,
$U = (U_x, U_y, U_z)$, along with the Lorentz Force's charge
velocity, v in $F = q(E + v \times B)$, will be the charge's velocity,
$C = (C_x, C_y, C_z)$ below.

Case 1 The charge is static, and it's velocity in s-frame is
$C = (0, 0, 0) = (C_x, C_y, C_z)$, $U_x = C_x = 0$, so
$C_{x'} = (0-v)/1 = -v$, and $U_y = C_y = 0$, so $C_{y'} = 0$, and
$U_z = C_z = 0$, so $C_{z'} = 0$
So $C' = (-v, 0, 0)$ same for the flux's velocity, $f' = (-v, 0, 0)$.

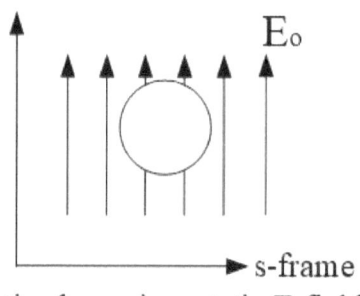 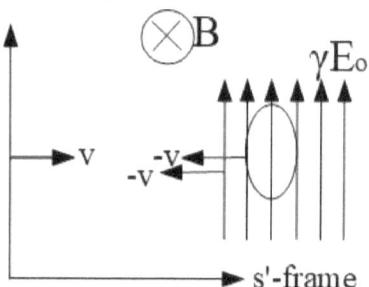

Static charge in a static **E** field   Both the charge and the flux
have a velocity of -v

For my pressure times area model I start with $F = qE_o$ and
look at it as the **E** field applying a pressure on a sphere, q, of radius
a. Thus I simply multiply and divide the right side by the area the
pressure is pushing on; $F = qE_o(\pi a^2/\pi a^2) = (qE_o/\pi a^2)(\pi a^2)$.

Note: In the rest frame of the electric field I'll call
$(qE_o/\pi a^2) = P_o = $ "Rest Pressure". Then, when flux lines are
crossing the charge, I will multiply the rest pressure by a "Pressure
Factor", $P_f = \gamma_v$, where v is the velocity of the flux w.r.t. the

charge as seen in an appropriate frame (See appendix).

Exploiting the fact that static pressure is an invariant scalar, I can get it from the charge's rest frame, which is $qE_o/\pi a^2$, and apply it to the s'-frame.

I see that the charge is a sphere length contracted into an ellipse

I see that the charge sees a vertically oriented field of flux traveling along with it.

The orientation of flux and ellipse means the the cross sectional area encountering the flux is an ellipse with semi major axis, a, sticking out of the page and semi minor axis of a/γ with normal in y-direction, so $\mathbf{A'} = (0, \pi a^2/\gamma, 0)$.

That the charge and flux are moving together means that the pressure factor $P_f = 1$. And the s-frame has a sphere of radius a, thus a cross section of πa^2, so

$$\mathbf{F} = (P_o)(P_f)(\mathbf{A}) = (qE_o/\pi a^2)(1)(0, \pi a^2, 0),$$
$$\mathbf{F} = (0, qE_o, 0)$$

And the s'-frame has the same sphere length contracted perpendicular to the charge's rest frame flux lines, thus a cross section of $\pi a^2/\gamma$, in the y direction, so $\mathbf{Area'} = (0, \pi a^2/\gamma, 0)$, and

$$\mathbf{F'} = (P_o)(P_f)(\mathbf{A'}) = (qE_o/\pi a^2)(1)(0, \pi a^2/\gamma, 0),$$
$$\mathbf{F'} = (0, qE_o/\gamma, 0)$$

Matching the traditional solutions.

Case 2 has the charge in the s-frame moving up parallel to **E** field

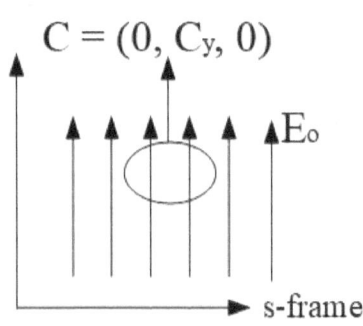 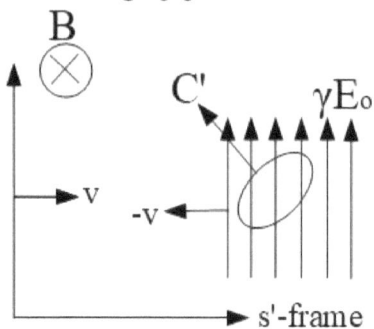

$C = (0, C_y, 0)$

E_o

s-frame

Charge moving parallel
to a static **E** field

B

C'

γE_o

v

-v

s'-frame

Both the charge and the flux
have same x'-velocity of -v

Look closer at the ellipse shown below between the vertical tangents. I think of it as a sail with a component of force perpendicular to the wind hitting it.

the **top** of the ellipse is experiencing a reinforced wind for a **lower** pressure, while

the **bottom** is experiencing a canceled wind for a **higher** pressure.

The line between the tangent points below is the Area cross section seen on edge. It is also an ellipse with semi minor axis from $(0, 0)$ to (x_o, y_o) and semi major axis, a, sticking straight out of the page.

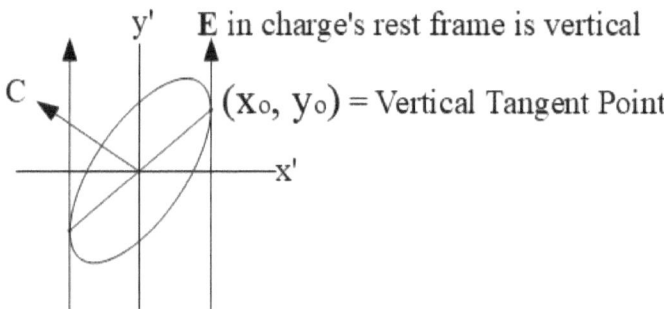

y' **E** in charge's rest frame is vertical

C

(x_o, y_o) = Vertical Tangent Point

-x'

A Tachyon Theory of Everything

By symmetry, the cross sections for both the x-directed force (between $+/- y_o$) and the y-directed force (between $+/-x_o$) are ellipses with semi major axis, a, directed straight out of the page, and the semi minor axis being either the x_o or y_o. For now I will say, "By inspection" F_x is negative and F_y is positive.

Since the area of an ellipse is πab where a is the semi major axis and b in the semi minor axis, need to solve for the vertical tangent point of an appropriately tilted ellipse.

Start with a sphere of radius, a, moving horizontally at speed C. Then $\beta_c = C/c$ and $\gamma_c = 1/\sqrt{1- \beta_c^2}$. This gives a vertically oriented ellipse with semi major axis, a, along the y-axis and semi minor axis, $b = a/\gamma_c$, along the x-axis.

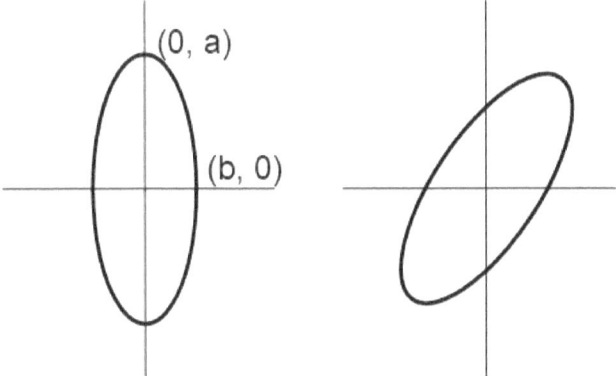

Dropping the subscripts for now and using double primes, ", to indicate rotated coordinates;

$$\frac{x^2}{b^2} + \frac{y^2}{a^2} = 1 \text{ , where } b = \frac{a}{\gamma} \text{ , } \frac{x^2}{\frac{a^2}{\gamma^2}} + \frac{y^2}{a^2} = 1 \text{ , } \gamma^2 x^2 + y^2 = a^2$$

Appendix Magnetic Force

Rotate equation by $-\theta$, or axis by $+\theta$

$$\begin{pmatrix} x \\ y \end{pmatrix} = \begin{pmatrix} \cos\theta & -\sin\theta \\ \sin\theta & \cos\theta \end{pmatrix}\begin{pmatrix} x'' \\ y'' \end{pmatrix} = \begin{array}{l} x = x''\cos\theta - y''\sin\theta \\ y = x''\sin\theta + y''\cos\theta \end{array}$$

Substitue the double prime variables into the equation;

$$\gamma^2(x''\cos\theta - y''\sin\theta)^2 + (x''\sin + y''\cos\theta)^2 = a^2$$

Dropping the double prime notation, skipping lots of algebra, and using the relation $\frac{1}{\gamma^2} = (1-\beta_c^2)$ for the charge's velocity;

$$x^2(1-\beta_c^2\sin^2\theta) - 2xy\beta_c^2\cos\theta\sin\theta + y^2(1-\beta_c^2\cos^2\theta) - (1-\beta_c^2)a^2 = 0$$

Solve for y using the quadratic equation;

$$y = \frac{--2x\beta_c^2\cos\theta\sin\theta}{2(1-\beta_c^2\cos^2\theta)} \pm$$

$$\frac{\sqrt{4x^2\beta_c^4\cos^2\theta\sin^2\theta - 4(1-\beta_c^2\cos^2\theta)(x^2(1-\beta_c^2\sin^2\theta) - (1-\beta_c^2)a^2)}}{2(1-\beta_c^2\cos^2\theta)}$$

Schematically, for the purposes of taking the derivative to find the vertical slope;

$$y = \frac{Ax \pm \sqrt{f(x)}}{B}, \quad \frac{dy}{dx} = \frac{1}{B}\left| A \pm \frac{1}{2}(f(x))^{-\frac{1}{2}}\frac{df}{dx} \right|$$

When the square root is zero, the slope is vertical. Setting the square root to zero, and solving for $X = X_o$;

$$0 = 4x^2\beta_c^4\cos^2\theta\sin^2\theta - 4(1-\beta_c^2\cos^2\theta)(x^2(1-\beta_c^2\sin^2\theta) - (1-\beta_c^2)a^2)$$

skipping some algebra;

$$X_o = \pm a\sqrt{1-\beta_c^2\cos^2\theta} = \text{semi minor axis of ellipse for } A_{y'},$$

the area of the cross section pointing in the $y-$direction.

Plug X_o into y to get the semi minor axis of the ellipse for, $A_{x'}$;

$$y = \frac{2\left(\pm a\sqrt{1-\beta_c^2\cos^2\theta}\right)\beta_c^2\cos\theta\sin\theta \pm \sqrt{0}}{2(1-\beta_c^2\cos^2\theta)} \quad,$$

Note for this x, the last, $\sqrt{\ }$, term is zero, and $y = y_o$.

$$y_o = \frac{\pm a\beta_c^2\cos\theta\sin\theta}{\sqrt{1-\beta_c^2\cos^2\theta}} \ , \ \beta_c = \frac{C'}{c} \ , \ \cos\theta = \frac{v}{C'} \ , \ \sin\theta = \frac{C_y'}{C'} \ , \ \gamma = \frac{1}{\sqrt{1-\frac{v^2}{c^2}}}$$

$$(x_o, y_o) = \left| a\sqrt{1-\beta_c^2\cos^2\theta}, \ \frac{a\beta_c^2\cos\theta\sin\theta}{\sqrt{1-\beta_c^2\cos^2\theta}} \right|$$

$$(x_o, y_o) = \left| a\sqrt{1-\left(\frac{C'^2}{c^2}\right)\left(\frac{v^2}{C'^2}\right)}, \ \frac{a\left(\frac{C'^2}{c^2}\right)\left(\frac{v}{C'}\right)\left(\frac{C_y'}{C'}\right)}{\sqrt{1-\left(\frac{C'^2}{c^2}\right)\left(\frac{v^2}{C'^2}\right)}} \right|$$

$$(x_o, y_o) = a\left(\frac{1}{\gamma}, \frac{v}{c}\frac{C_y'}{c}\gamma\right), \text{ since } \gamma C_y' = C_y \ , \ (x_o, y_o) = a\left(\frac{1}{\gamma}, \frac{v}{c}\frac{C_y}{c}\right)$$

Then getting the cross sectional area vector;

$$A_x' = -\pi a y_o = -\pi a^2\frac{vC_y}{c^2} \ , \ A_y' = \pi a x_o = \pi\frac{a^2}{\gamma} \ , \ \vec{A'} = \pi a^2\left(\frac{-vC_y}{c^2}, \frac{1}{\gamma}, 0\right)$$

Since the charge is not crossing flux lines, $P_f = 1$, and

$$\mathbf{F} = (P_o)(P_f)(\mathbf{A'})$$

$$\vec{F'} = \left(\frac{qE_o}{\pi a^2}\right)(1)\pi a^2\left(\frac{-vC_y}{c^2}, \frac{1}{\gamma}, 0\right) = \left(\frac{-qE_o vC_y}{c^2}, \frac{qE_o}{\gamma}, 0\right)$$

Matching the traditional solutions.

Case 3 has the charge moving horizontally at velocity,v, in the
s-frame. So $U_x = C_x = v$

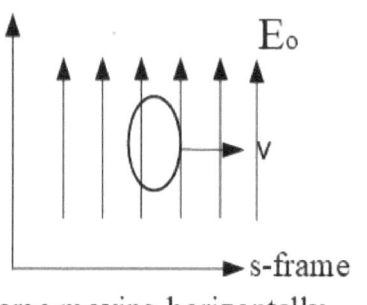

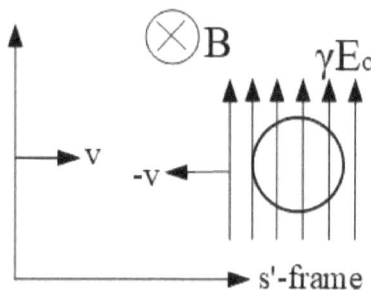

Charge moving horizontally
in a static **E** field

Charge is stationary while
the flux has a velocity of -v

    For the Pressure times area argument the moving charge in
the s-frame has the length contracted cross sectional area of $\pi a^2/\gamma$
with the vector pointing straight up, so $\mathbf{A} = (0, \pi a^2/\gamma, 0)$.
The charge sees a moving and thus length contracted and thus
concentrated flux field, so, $P_f = \gamma_v$, where $v = C_x$, or the usual γ.
$\mathbf{F} = (P_o)(P_f)\mathbf{A} = (qE_o/\pi a^2)(\gamma)(0, \pi a^2/\gamma, 0)$ Multiplying them
together cancels everything but the qE_o. So $\mathbf{F} = (0, qE_o, 0)$
    Then in the s'-frame the same pressure is acting on the full
cross section of the sphere, $\mathbf{A'} = (0, \pi a^2, 0)$, so the gamma is not
canceled, and

$$\mathbf{F'} = (0, \gamma qE_o, 0)$$

Matching the traditional solutions.

Case 4 s-frame: The charge going up at an angle.

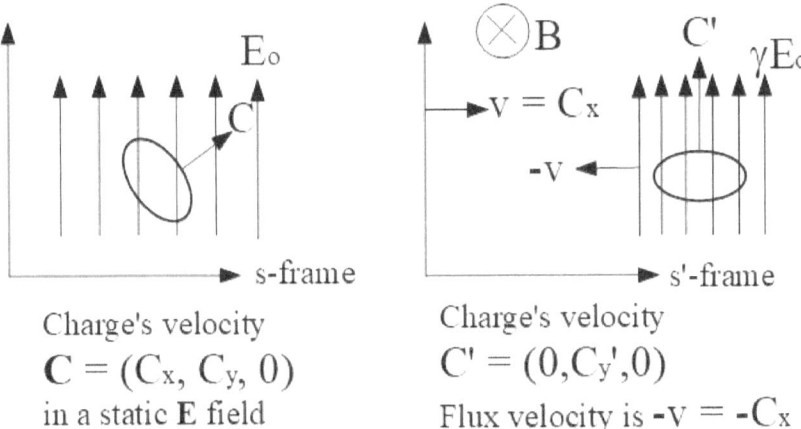

Charge's velocity

$$C = (C_x, C_y, 0)$$

in a static **E** field

Charge's velocity

$$C' = (0, C_y', 0)$$

Flux velocity is $-v = -C_x$

For the s-frame, $F = (0, qE_o, 0)$, but the pressure times area model appears to not work since the ellipse looks like it should have a force to the right. Thus introducing the rule;

When flux lines are crossing the charge, and the charge has a component of velocity parallel to the flux, the charge observes an aberrated flux field in the pressure times area model. See the appendix on Aberration.

A note on notation. From now on **all velocities are β's**, and a subscript after a β or a γ means that subscript is the velocity of interest in the function. So above, what was $C_x/c = \beta_{Cx}$ is now $\beta_{Cx} = C_x$, similarly $\beta_{Cy} = C_y$ and $\beta_C = C$ = speed of charge, and $\gamma_c = 1/\sqrt{1 - \beta_c^2}$ but a y subscript means C_y, the charges y-velocity (again, w.r.t. c), and but lower case c is still light speed everywhere. Most of the velocities in the transforms from Resnick are already divided by c, those that are not can be made β's by simply either dividing through the whole equation by c, or multiplying the appropriate velocity by c/c. The electric field transformations look nicer if the stray c is put next to the **B**, same for the Lorentz force, $\mathbf{F} = q(\mathbf{E} + \boldsymbol{\beta} \times c\mathbf{B})$. Indeed it will be convenient to keep c next to **B** all of the time. Although I will revert to β_c, etc., in a few pages but only for a few pages.

Equate the slope of the **E** field which is rise over run (take the y-component and divide by the x-component) to the slope or derivative of the ellipse representing the charge in the s-frame to get the cross sectional area. But I have to make sure I am using variables that are all in the same frame.

For the aberration, by "see what the charge sees", I mean In my frame I have the infinite plane's velocity, **P,** telling me how fast I see the flux moving across the surface of the charge. Now, focusing on the speed of the flux across the surface, I transform to the "rest frame of the velocity difference" to then get the velocity difference at the charge, then transform this velocity difference back to my frame. If, when transforming to the rest frame of the velocity difference, you are in the rest frame of either the charge or the flux, then the velocity difference is just the other object's velocity and transforming up and back just gets back the same velocity.

First, I observe the flux or Plane of source charge,**P** , must be moving at - C_x. Thus I see that the flux is moving across the surface of the charge at - C_x. Transforming to the rest frame of the velocity difference, I am "catching up" to the charge with a boost of C_y, and calculate the velocity difference in the rest frame of the velocity difference to be $\gamma_y(- C_x)$, but transforming back to my frame means dividing by the same γ_y leaving the original $(- C_x)$. Also, the boost of C_y to the rest frame of the velocity difference also means that the charge sees the source plane moving down at $- C_y$, so "I see that the charge sees" the source plane moving at

$(- C_x, - C_y, 0)$, so $\beta\cos\theta = - C_x$ and $\beta\sin\theta = - C_y$ in the formula.

$$E_{Aberrated} = \frac{\sigma''}{2\varepsilon_o}\left(\frac{-\beta\cos\theta\beta\sin\theta}{1-\beta^2\sin^2\theta}, 1, 0\right) = \frac{\sigma''}{2\varepsilon_o}\left|\frac{-C_xC_y}{1-C_y^2}, 1, 0\right|$$

$$E_{Aberrated} = \frac{\sigma''}{2\varepsilon_o}\left(-\gamma_y^2 C_x C_y, 1, 0\right)$$

$$Slope = \frac{rise}{run} = \frac{1}{-\gamma_y^2 C_x C_y} = \frac{-1}{\gamma_y^2 C_x C_y}$$

Now, the slope of this field (y-component divided by x-component) is set equal to the slope from the derivative to get the cross sectional area that the pressure is acting on.

The observed ellipse is rotated by a negative theta w.r.t. the ellipse described in Case 2. Since the only non-squared trig functions in the equation for the tilted ellipse are the $\cos\theta\sin\theta$ out in front of the square root sign, I can get my desired equation by just flipping the sign in front of that term. Thus;

$$y = \frac{-2x\beta_c^2\cos\theta\sin\theta}{2(1-\beta_c^2\cos^2\theta)} \pm$$

$$\frac{\sqrt{4x^2\beta_c^4\cos^2\theta\sin^2\theta - 4(1-\beta_c^2\cos^2\theta)(x^2(1-\beta_c^2\sin^2\theta)-(1-\beta_c^2)a^2)}}{2(1-\beta_c^2\cos^2\theta)}$$

$$y = \frac{-x\beta_c^2\cos\theta\sin\theta \pm \sqrt{1-\beta_c^2}\sqrt{(1-\beta_c^2\cos^2\theta)a^2 - x^2}}{(1-\beta_c^2\cos^2\theta)}$$

Now take the derivative w.r.t. x, but let the denominator equal K, and pull it out in front;

$$\frac{dy}{dx} = \frac{1}{K}\left[-\beta_c^2\cos\theta\sin\theta \pm \left(\sqrt{1-\beta_c^2}\right)\frac{1}{2}\left(a^2(1-\beta_c^2\cos^2\theta)-x^2\right)^{-\frac{1}{2}}(0-2x)\right]$$

$$\frac{dy}{dx} = \frac{1}{(1-\beta_c^2\cos^2\theta)}\left[-\beta_c^2\cos\theta\sin\theta \pm -\frac{x\sqrt{1-\beta_c^2}}{\sqrt{a^2(1-\beta_c^2\cos^2\theta)-x^2}}\right]$$

In this case the observed ellipse is the charge in the s-frame, so $\beta_c\cos\theta$ and $\beta_c\sin\theta$ are just the x, and y components of the charge's velocity. To avoid too many subscripts when equating the above derivative to the slope of the aberrated field set
$C = \beta_c$, $C_x = \beta_c\cos\theta$, $C_y = \beta_c\sin\theta$

$$\frac{dy}{dx} = \frac{-1}{(1-C_x^2)}\left[C_xC_y \pm \frac{x\sqrt{1-C^2}}{\sqrt{a^2(1-C_x^2)-x^2}}\right] = \frac{-1}{C_xC_y\gamma_y^2} = \begin{pmatrix}\text{Slope of}\\ E_{\text{Aberrated}}\end{pmatrix}$$

Skipping some algebra on the way to solving for x;

$$x^2 = a^2\frac{(1-C^2)}{(1-C_y^2)} = \frac{a^2\gamma_y^2}{\gamma_c^2} \quad , \quad x = \pm\frac{a\gamma_y}{\gamma_c}$$

x is the semi minor axis of the cross sectional ellipse pointed in the y-direction, So the cross sectional area, oriented along the plus y-direction is an ellipse with semi major axis, "a" (sticking out of the page), and semi minor axis $x = a\gamma_y/\gamma_c$. Thus $A_y = \pi a^2\gamma_y/\gamma_c$.

To get A_x, plug the just solved for x value into the ellipse equation to solve for y, the semi minor axis of the ellipse that is the cross sectional area along the x-axis;

$$y = \frac{-xC_xC_y \pm \sqrt{(1-C^2)}\sqrt{a^2(1-C_x^2) - x^2}}{1 - C_x^2}.$$

Plugging in $x = \dfrac{a\gamma_y}{\gamma_c} = \dfrac{a\sqrt{1-C^2}}{\sqrt{1-C_y^2}}$

and using the positive case from the $\pm$ results in $y = 0$.

Thus $A_x = 0$, and the area vector, $\mathbf{A} = (0, \pi a^2 \gamma_y/\gamma_c, 0)$

For the pressure factor, I need the σ'' the charge sees. I know it sees a field concentrated by γ_v, where v is the velocity the charge sees the flux lines moving at. Thus, as explained in the appendix, I transform to the "Rest frame of the velocity difference" with a boost, parallel to the static $\mathbf{E}$ field, of $\mathbf{v}' = (0, C_y, 0)$ from the s-frame. Since the s-frame already shows static flux lines, boosting parallel to $\mathbf{E}$ keeps the flux lines stationary and the velocity difference is just the charge's velocity (as explained in the appendix) that we have "Caught up to" (as explained in the appendix). Thus where the s-frame sees a charge flux velocity difference as C_x, I know the charge sees a flux velocity of $\gamma_y C_x$, so this is what I plug into the γ function to concentrate the flux lines;

$$\sigma'' = \frac{q}{\Delta x'' \Delta z''}, \text{ plugging in } \Delta x'' = \frac{\Delta x}{\gamma_{(\gamma_y C_x)}} \text{ and } \Delta z'' = \Delta z$$

$$\sigma'' = \frac{q}{\dfrac{\Delta x}{\gamma_{(\gamma_y C_x)}}\Delta z} = \sigma\gamma_{(\gamma_y C_x)} = \sigma\frac{1}{\sqrt{1 - (\gamma_y C_x)^2}} = \frac{\sigma}{\sqrt{1 - \left(\dfrac{C_x}{\sqrt{1-C_y^2}}\right)^2}}$$

Appendix Magnetic Force

$$\sigma'' = \sigma\sqrt{\frac{1 - C_y^2}{1 - \left(C_x^2 + C_y^2\right)}} = \sigma\frac{\gamma_c}{\gamma_y}$$

$\dfrac{\gamma_c}{\gamma_y}$ is the "Pressure Factor", P_f , while $\dfrac{\sigma q}{2\varepsilon\pi a^2}$ is $P_o = \dfrac{qE_o}{\pi a^2}$,

with $E_o = \dfrac{\sigma}{2\varepsilon}$ being the s $-$ frame electric field.

$$\vec{F} = P_o P_f \overrightarrow{Area} = \frac{\sigma q}{2\varepsilon\pi a^2}\frac{\gamma_c}{\gamma_y}\left(0,\ \pi a^2\frac{\gamma_y}{\gamma_c},\ 0\right) = (0,\ qE_o,\ 0)$$

Matching the traditional answer.

Wow, a whole lot of work for a very simple answer. To review, I needed three things, 1 the speed of the flux the responding charge sees for the pressure factor, 2 that same speed difference transformed back to my frame along with 3 the shape of the responding charge that I see (Usually a tilted ellipse) to get the tangents. A geometrical interpretation that I will apply consistently in the next several cases.

Now to continue with **Case 4 s'-frame** and it's magnetic field. Referencing the same picture but looking at the s'-frame; $C = (C_x,\ C_y,\ 0)$, and because the s'-frame is "Catching up" to the C_y velocity with a boost of C_x, $C' = (0,\ \gamma_x C_y,\ 0)$.

I see the charge as a horizontally oriented ellipse with semi minor axis, b, length contracted by γ_y '.

$$\frac{x^2}{a^2} + \frac{y^2}{b^2} = 1 \ , \ b = \frac{a}{\gamma_{y'}} \ , \ x^2 + \gamma_{y'}^2 y^2 = a^2 \ , \text{or} \ y = \frac{1}{\gamma_{y'}}\left(a^2 - x^2\right)^{\frac{1}{2}}$$

$$\frac{dy}{dx} = \frac{1}{\gamma_{y'}}\frac{1}{2}\left(a^2 - x^2\right)^{-\frac{1}{2}}(0 - 2x) = \frac{-x}{\gamma_{y'}\sqrt{a^2 - x^2}}$$

For the aberration I concentrate on the flux. In my frame I have **p**, the infinite plane of source charge's velocity, telling me how fast I see the flux moving across the surface of the charge. Now, focusing on the speed of the flux across the surface, I transform to the "rest frame of the velocity difference" to then get the velocity difference at the charge, then transform this velocity difference back to my frame. If, when transforming to the rest frame of the velocity difference, you are in the rest frame of either the charge or the flux, then the velocity difference is just the other object's velocity and transforming up and back just gets back the same velocity.

A little aside about what is being "caught up" to or "run away" from. As implied in the Aberration appendix, the uniform **E** field is produced by an infinite plane of charge right under foot. In the last case, the flux was still and a vertical boost keeps it still, but the charge's x-velocity is "caught up" to. Thus the velocity difference is a "caught up" velocity. Here the charge has no x-velocity and a vertical boost keeps it zero. But the plane of flux under foot is being "run away" from, so the velocity difference is a "run away" velocity.

First, I observe the flux or Plane of source charge, **p**, is moving at $-C_x$. Thus I see that the flux is moving across the surface of the charge at $-C_x$. Transforming to the rest frame of the velocity difference, I am "running away" from the plane with a

Appendix Magnetic Force

boost of C_y', and calculate the velocity difference in the rest frame of the velocity difference to be $(- C_x)/\gamma_y$' (This is the Cx" in the diagram on the next page.), but transforming back to my frame means multiplying by the same γ_y' leaving the original $(- C_x)$. Also, the boost of C_y' to the rest frame of the velocity difference means the charge sees the source plane moving down at $- C_y$'. So "I see that the charge sees" a flux velocity of $\mathbf{f} = (- C_x, - C_y', 0)$ Thus I plug in $\beta\cos\theta = - C_x$ and $\beta\sin\theta = - C_y$' into the Aberration formula and equate the derivative of the ellipse to the slope (y-component/x-component) of the aberrated field;

$$\frac{dy}{dx} = \frac{-x}{\gamma_{y'}\sqrt{a^2 - x^2}} = \frac{-1}{\gamma_{y'}^2 C_{y'} C_x} = \text{Slope of } E_{\text{Aberrated}}$$

Solve for x to get y $-$ oriented semi minor axis

$$x = \pm\frac{a}{\sqrt{\gamma_{y'}^2 C_{y'}^2 C_x^2 + 1}} \quad , \quad \text{Thus } A_y = \frac{\pi a^2}{\sqrt{\gamma_{y'}^2 C_{y'}^2 C_x^2 + 1}}$$

Plugging the x back into ellipse equation to get y;

$$y = \frac{1}{\gamma_{y'}}\sqrt{a^2 - \left(\frac{a}{\sqrt{\gamma_{y'}^2 C_{y'}^2 C_x^2 + 1}}\right)^2} = \frac{aC_x C_{y'}}{\sqrt{\gamma_{y'}^2 C_{y'}^2 C_x^2 + 1}}$$

And $A_x = \dfrac{\pi a^2 C_x C_{y'}}{\sqrt{\gamma_{y'}^2 C_{y'}^2 C_x^2 + 1}}$, So $\vec{A} = \dfrac{\pi a^2}{\sqrt{\gamma_{y'}^2 C_{y'}^2 C_x^2 + 1}}(-C_x C_{y'}, 1, 0)$

For the pressure factor, Now the flux is moving while the charge is still, so I look at the source plane's velocity. I use the following geometry, with the ellipse being one charge in the plane as seen by the responding charge. To get the perpendicular flux velocity, I use the velocity of the source plane as seen by the charge, so again, transforming to the "Rest Frame of the Velocity Difference" between the flux and charge, I get the flux velocity used in the pressure factor by looking at the perpendicular velocity of the flux lines.

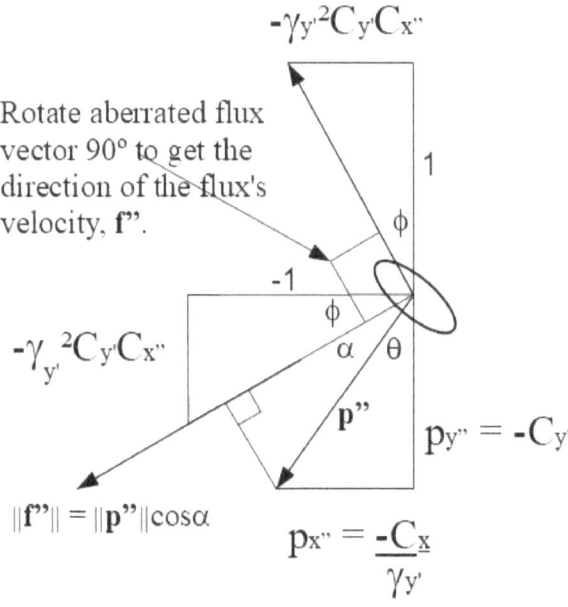

$$-\gamma_{y'}^2 C_{y'} C_{x''}$$

Rotate aberrated flux vector 90° to get the direction of the flux's velocity, **f''**.

$$-\gamma_{y'}^2 C_{y'} C_{x''}$$

$$\|\mathbf{f''}\| = \|\mathbf{p''}\|\cos\alpha$$

$$p_{y''} = -C_{y'}$$

$$p_{x''} = \frac{-C_x}{\gamma_{y'}}$$

The boost velocity to the rest frame of the velocity difference between the flux and charge is $\mathbf{v'} = (0, C_{y'}, 0)$, so "running away" as opposed to "catching up" to the plane of the source charge, so

$p_{x''} = p_x/\gamma_{y'} = -C_x/\gamma_{y'} = C_{x''}$ for pluging into the Pressure factor.

$p_{y''} = -v' = -C_{y'} = -\gamma_x C_y$

$$f'' = p''\cos\alpha = p''\cos\left(\frac{\pi}{2} - (\theta + \varphi)\right) = p''\sin(\theta + \varphi)$$

$$f'' = p''(\sin\theta\cos\varphi + \cos\theta\sin\varphi)$$

Plugging in the trig functions from the diagram;

$$f'' = p''\left(\left(\frac{-\frac{C_x}{\gamma_{y'}}}{p''}\right)\left(\frac{-1}{\sqrt{\gamma_{y'}^4 C_{y'}^2 C_{x''}^2 + 1}}\right) + \left(\frac{-C_{y'}}{p''}\right)\left(\frac{-\gamma_{y'}^2 C_y C_{x''}}{\sqrt{\gamma_{y'}^4 C_{y'}^2 C_{x''}^2 + 1}}\right)\right)$$

Using the relation, $C_{x''} = \dfrac{C_x}{\gamma_{y'}} = C_x\sqrt{1 - C_{y'}}$, and simplifying;

$$f'' = \frac{C_x}{\sqrt{C_x^2 C_{y'}^2 + 1 - C_{y'}^2}}$$

This flux velocity from the "Rest frame of the velocity difference" is what I plug into the gamma function to get $\gamma_{f''} = P_f$, the pressure factor. Plugging the flux velocity into a gamma function gives $\gamma_{f''} = P_f$, the pressure factor;

$$P_f = \gamma_{f''} = \frac{1}{\sqrt{1 - \left(\frac{C_x}{\sqrt{C_x^2 C_{y'}^2 + 1 - C_{y'}^2}}\right)^2}} = \gamma_x\sqrt{\gamma_{y'}^2 C_{y'}^2 C_x^2 + 1}$$

Now that we have everything for the force;

$$\vec{F} = (P_o)(P_f)\left(\overrightarrow{Area}\right)$$

$$\vec{F} = \left(\frac{qE_o}{\pi a^2}\right)\left(\gamma_x\sqrt{\gamma_{y'}^2 C_y^2 C_x^2 + 1}\right)\left|\frac{\pi a^2}{\sqrt{\gamma_{y'}^2 C_y^2 C_x^2 + 1}}(-C_x C_{y'}, 1, 0)\right|$$

Using $\gamma_x C_y = C_{y'}$, and simplifying; $\vec{F} = qE_o\gamma_x(-C_x\gamma_x C_y, 1, 0)$

Matching the traditional answers.

Case 4 s"-frame, the charge's rest frame, Boosting to charge's rest frame with $v' = (0, C_{y'}, 0)$, has the observer and the charge already in the invariant velocity difference set of frames, so no "Up and Back" logic since we are already there. So I just plug in the observed $P_{x"} = C_{x'}/\gamma_{y'} = \beta\cos\theta$ for the flux velocity. Since it is now acting on a circle, the direction of the force equals the direction of the aberration, and, I've already calculated the pressure factor so;

$$x^2 + y^2 = a^2 \text{ or } y = \left(a^2 - x^2\right)^{\frac{1}{2}}$$

$$\frac{dy}{dx} = \frac{1}{2}\left(a^2 - x^2\right)^{-\frac{1}{2}}(0 - 2x) = \frac{-x}{\sqrt{a^2 - x^2}}$$

$$\frac{dy}{dx} = \frac{-x}{\sqrt{a^2 - x^2}} = \frac{-1}{\gamma_{y'}^2 C_{x"} C_{y'}} = \text{Slope of } E_{\text{Aberrated}}$$

Using $C_{x"} = \frac{C_{x'}}{\gamma_{y'}}$, and solving for x gives

$$\frac{\pm a}{\sqrt{\gamma_{y'}^2 C_x^2 C_{y'}^2 + 1}}y = x, \text{ semi minor axis for } A_y.$$

Appendix Magnetic Force

Plugging back into y;

$$y = \sqrt{a^2 - \left(\frac{a}{\sqrt{\gamma_{y'}^2 C_x^2 C_{y'}^2 + 1}}\right)^2}$$

$$y = \frac{a\gamma_{y'} C_x C_{y'}}{\sqrt{\gamma_{y'}^2 C_x^2 C_{y'}^2 + 1}} = \text{semi minor axis for } A_x, \text{ thus}$$

$$\overrightarrow{\text{Area}} = \frac{\pi a^2}{\sqrt{\gamma_{y'}^2 C_x^2 C_{y'}^2 + 1}} \left(-\gamma_{y'} C_x C_{y'}, 1, 0\right)$$

Calculating the force using the same pressure factor as before;

$$\vec{F} = (P_o)(P_f)\left(\overrightarrow{\text{Area}}\right)$$

$$\vec{F} = \left(\frac{qE_o}{\pi a^2}\right)\left(\gamma_x \sqrt{\gamma_{y'}^2 C_x^2 C_{y'}^2 + 1}\right)\left|\frac{\pi a^2}{\sqrt{\gamma_{y'}^2 C_x^2 C_{y'}^2 + 1}}\left(-\gamma_{y'} C_x C_{y'}, 1, 0\right)\right|$$

$$\vec{F} = qE_o \gamma_x \left(-\gamma_{y'} C_x C_{y'}, 1, 0\right)$$

Matching the traditional answers.

A Tachyon Theory of Everything

Case 5 is the s'-frame with and arbitrary boost so both the charge and flux have an x-velocity

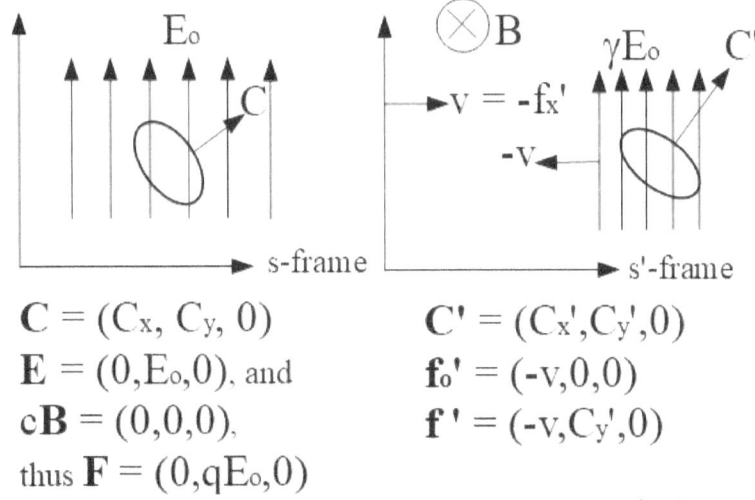

$$C = (C_x, C_y, 0)$$
$$E = (0, E_o, 0), \text{ and}$$
$$cB = (0, 0, 0),$$
thus $F = (0, qE_o, 0)$

$$C' = (C_x', C_y', 0)$$
$$f_o' = (-v, 0, 0)$$
$$f' = (-v, C_y', 0)$$

It is in this case that I use the whole invariant velocity difference with "catch up and back" logic described after aberration, in the aberration formula. Also same ellipse and it's derivative as case 4.

Everything is in s', so drop primes for now

$$\frac{dy}{dx} = \frac{-1}{1-C_x^2}\left[C_xC_y \pm \frac{x\sqrt{1-C^2}}{\sqrt{a^2(1-C_x^2)-x^2}}\right] = \frac{-1}{\gamma_y^2\left(\dfrac{C_x-f_x}{1-\gamma_y^2 f_x C_x}\right)C_y}$$

Notice the "Invariant Velocity Difference" of, $\dfrac{C_x - f_x}{1 - \gamma_y^2 f_x C_x}$,

substituted in for C_x in the formula for the slope of $E_{\text{aberrated}}$
Skipping lots of algebra, and solving for "x" to get the semi-minor axis of A_y;

$$x = a\sqrt{\frac{\left(1 - C_x^2 - C_y^2\right)\left(1 - f_xC_x\right)^2}{\left(1 - f_xC_x\right)^2 - C_y^2\left(1 - f_x^2\right)}} \text{ , and}$$

$$A_y = \pi a^2 \sqrt{\frac{\left(1 - C_x^2 - C_y^2\right)\left(1 - f_xC_x\right)^2}{\left(1 - f_xC_x\right)^2 - C_y^2\left(1 - f_x^2\right)}}$$

Plugging this x into $y = \dfrac{-C_xC_yx \pm \sqrt{1 - C^2}\sqrt{a^2\left(1 - C_x^2\right) - x^2}}{1 - C_x^2}$,

and using $C^2 = C_x^2 + C_y^2$ I get after lots more algebra;

$$y = \frac{-af_xC_y\sqrt{1 - C_x^2 - C_y^2}}{\sqrt{\left(1 - f_xC_x\right)^2 - C_y^2\left(1 - f_x^2\right)}} \text{ , So}$$

$$\overrightarrow{\text{Area}} = \frac{\pi a^2\sqrt{1 - C_x^2 - C_y^2}}{\sqrt{\left(1 - f_xC_x\right)^2 - C_y^2\left(1 - f_x^2\right)}}\left(f_xC_y,\ 1 - f_xC_x,\ 0\right)$$

The y looks negative, but the f_x is also negative, so the tangent point is in the first quadrant, and therefore the plane defined by the tangents, going through zero, must be tilted in the plus theta direction thus **Area-x** is negative with the negative f_x in the vector components.

Continuing with the pressure factor transformed in the appendix;

$$P_f = \sqrt{\frac{\left(1 - f_xC_x\right)^2 - C_y^2\left(1 - f_x^2\right)}{\left(1 - C_x^2 - C_y^2\right)\left(1 - f_x^2\right)}}$$

A Tachyon Theory of Everything

With the **Area** vector and P_f, we're ready to plug in;

$$\vec{F} = (P_o)(P_f)\left(\overrightarrow{\text{Area}}\right)$$

$$\vec{F} = \left(\frac{qE_o}{\pi a^2}\right)\left[\sqrt{\frac{(1 - f_x C_x)^2 - C_y^2(1 - f_x^2)}{(1 - C_x^2 - C_y^2)(1 - f_x^2)}}\right] \times$$

$$\left[\pi a^2 \sqrt{\frac{1 - C_x^2 - C_y^2}{(1 - f_x C_x)^2 - C_y^2(1 - f_x^2)}}(f_x C_y, \; 1 - f_x C_x, \; 0)\right]$$

Simplifying, and putting back the primes;

$$\vec{F} = \left(\frac{qE_o}{\sqrt{1 - f_{x'}^2}}\right)(f_{x'}C_{y'}, \; 1 - f_{x'}C_{x'}, \; 0)$$

Matching the traditional answers.

Case 6 is the most general case by taking the last s'-frame from Case 5 and giving it a boost of $v' = (0, 0, -C_{z''})$ to the s''-frame. So both the charge and the flux have the same C_z component.

    As Case 2 shows, if flux lines are not being crossed, then there is no aberration. So the aberration remains in the (x,y)-plane, but the ellipsoid (a sphere length contracted along it's velocity) is tilted somewhat.

    Not wanting to go through all the algebra of the last problem, I tried going to vectors. I present the vector formulas on the next couple of pages without proof.

    In the following diagram, the page is the plane defined by the responding charge's velocity vector, $\mathbf{C}$, and the aberrated field vector, $\mathbf{E_A}$, then, rotate in the plane so the aberrated field is pointing straight up (vertical tangents are easier to solve for).

    Realizing that the cross sectional area defined by the tangents of the field to the ellipsoid is always an ellipse passing through the origin, and pressure always acts normal to a plane, I use the area of the ellipse and it's unit normal vector.

    In the diagram below the vector, $\mathbf{C} = (C_x, C_y, C_z)$, is the charge's velocity (w.r.t. c), The vector, $\mathbf{n}$, is the unit normal vector to the plane defined by the origin, the vertical tangent point (x_o, y_o), and the z-axis pointing straight out of the page. The vector $\mathbf{E_A}$ is the aberrated electric field gradient direction pointing straight up. The letter, a, is the radius of the charge and the semi major axis of the ellipse, while the letter, b, is the semi minor axis of the ellipse. θ, is the angle between the line going through the origin and the point (x_o, y_o), and the horizontal-axis.

A Tachyon Theory of Everything

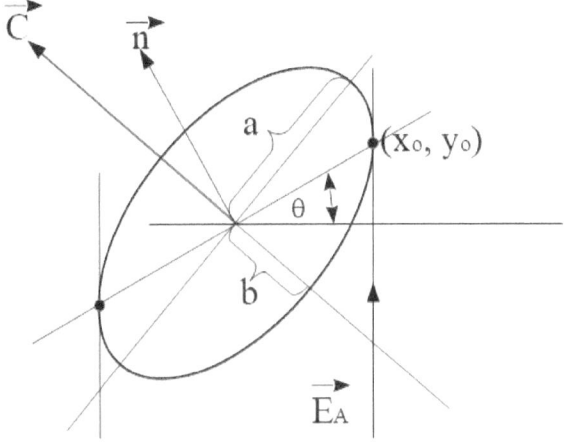

Note that the force on the charge will be directed along the normal vector, **n**. The **Area** vector is the Area scalar times the normal vector, **n**. The vector, **E**, (No subscript) is the static electric field as seen by a stationary observer (Magnitude is E_o). Non-bold, or non-vector notated variables (The variable missing the vector arrow.) are magnitudes. The vector, $\mathbf{f_o}$, is the flux velocity perpendicular to the flux lines. And an upward pointing carrot symbol above a variable indicates a unit vector.

$$\vec{f} = \vec{f_o} + \left(\vec{C} \cdot \hat{E}\right)\hat{E} \ , \ \ \vec{E_A} = \hat{E} + \frac{\left(\vec{f} - \vec{C}\right)\left(\vec{C} \cdot \hat{E}\right)}{1 - \vec{f} \cdot \vec{C}} \ \ ,$$

$$E_A = \sqrt{1 + \frac{\left(\vec{f} - \vec{C}\right) \cdot \left(\vec{f} - \vec{C}\right)\left(\vec{C} \cdot \hat{E}\right)^2}{\left(1 - \vec{f} \cdot \vec{C}\right)^2}}$$

Appendix Magnetic Force

$$x_o = \sqrt{a^2\cos^2\theta + b^2\sin^2\theta}, \quad y_o = \frac{\sin\theta\cos\theta(a^2 - b^2)}{\sqrt{a^2\cos^2\theta + b^2\sin^2\theta}}, \quad b^2 = a^2\left(1 - \vec{C}\cdot\vec{C}\right)$$

$$P_o = \frac{qE_o}{\pi a^2} = \text{Pressure on stationary charge in static electric field}$$

$$\hat{n} = \frac{\widehat{E_A}\left(1 - \vec{C}\cdot\vec{C}\right) + \vec{C}\left(\vec{C}\cdot\widehat{E_A}\right)}{\sqrt{\left(1 - \vec{C}\cdot\vec{C}\right)^2 + \left(2 - \vec{C}\cdot\vec{C}\right)\left(\vec{C}\cdot\widehat{E_A}\right)^2}} \;,$$

$$\text{Area} = \pi a^2 \sqrt{\frac{\left(1 - \vec{C}\cdot\vec{C}\right)^2 + \left(2 - \vec{C}\cdot\vec{C}\right)\left(\vec{C}\cdot\widehat{E_A}\right)^2}{1 - \vec{C}\cdot\vec{C} + \left(\vec{C}\cdot\widehat{E_A}\right)^2}}$$

Multiplying the above two together gets the **Area** vector;

$$\overrightarrow{\text{Area}} = \vec{n}(\text{Area}) = \pi a^2 \frac{\widehat{E_A}\left(1 - \vec{C}\cdot\vec{C}\right) + \vec{C}\left(\vec{C}\cdot\widehat{E_A}\right)}{\sqrt{1 - \vec{C}\cdot\vec{C} + \left(\vec{C}\cdot\widehat{E_A}\right)^2}}$$

And with the Pressure Factor from the appendix;

$$P_f = \sqrt{\frac{\left(1 - \vec{f}\cdot\vec{C}\right)^2 + \left(\vec{f} - \vec{C}\right)\cdot\left(\vec{f} - \vec{C}\right)\left(\vec{C}\cdot\hat{E}\right)^2 + \left(1 - \vec{C}\cdot\vec{C}\right)\left(\vec{C}\cdot\hat{E}\right)^2}{\left(1 - \vec{C}\cdot\vec{C}\right)\left(1 - \vec{f}\cdot\vec{f} + \left(\vec{C}\cdot\hat{E}\right)^2\right)}}$$

Putting them all together;

$$\vec{F} = (P_o)(P_f)\left(\overrightarrow{\text{Area}}\right) = qE_o \frac{\left(\left(1 - \vec{f}\cdot\vec{C}\right)\hat{E} + \left(\vec{C}\cdot\hat{E}\right)\vec{f}\right)}{\sqrt{1 - \vec{f}\cdot\vec{f} + \left(\vec{C}\cdot\hat{E}\right)^2}}$$

A Tachyon Theory of Everything

Notice the last equation from multiplying all the terms together and simplifying. With overline in type signifying unit vectors, Subbing $\overline{E} \rightarrow r$, $q f_o \rightarrow I$ up top while $f_o \rightarrow v$ down below (r = unit radius vector from source charge and current element to responding charge. I is the current element. And v is the velocity of the static E field) will yield the force from the transformed perpendicular E field.

Anyway, for this case;

$$\vec{C''} = \left(C_{x''},\, C_{y''},\, C_{z''}\right) \ , \ \vec{f''} = \left(f_{x''},\, C_{y''},\, C_{z''}\right) \ , \ \hat{E''} = (0,\, 1,\, 0)$$

$$\vec{C''} \cdot \hat{E''} = C_{y''} \ , \ \vec{f''} \cdot \vec{C''} = f_{x''} C_{x''} + C_{y''}^2 + C_{z''}^2 \ , \ \vec{f''} \cdot \vec{f''} = f_{x''}^2 + C_{y''}^2 + C_{z''}^2$$

Since everything is in the same double prime system, drop the double primes for now;

$$\vec{F} = qE_o \frac{\left(\hat{E}\left(1 - \vec{f} \cdot \vec{C}\right) + \vec{f}\left(\vec{C} \cdot \hat{E}\right)\right)}{\sqrt{1 - \vec{f} \cdot \vec{f} + \left(\vec{C} \cdot \hat{E}\right)^2}}$$

Plugging in the above;

$$\vec{F} = qE_o \frac{\left((0,\, 1,\, 0)\left(1 - f_x C_x - C_y^2 - C_z^2\right) + \left(f_x,\, C_y,\, C_z\right)C_y\right)}{\sqrt{1 - f_x^2 - C_y^2 - C_z^2 + C_y^2}}$$

$$\vec{F} = qE_o \frac{\left(f_x C_y,\, 1 - f_x C_x - C_z^2,\, C_y C_z\right)}{\sqrt{1 - f_x^2 - C_z^2}}$$

Adding back the double primes will match the traditional answers.

So, I have satisfied myself that the magnetic force is the un-canceled (Case 4 s-frame has the ellipse and aberration effects canceling) effects of relativity on the electric field's gradient and charge's sphere (The length contracted concentration of flux lines is taken care of with the gamma out front of the E and B transformation equations.).

The whole "see what the charge sees" thing is defended with the notion that a permanent magnet, the thing we used to discover and analyze magnetic fields, is basically a magnetic moment. Which is a fairly sophisticated device. As the charge travels in a circle, at any moment it is probing for any fields (distant charges) moving parallel or anti-parallel to it. By traveling in a circle it "looks at" all velocities in the plane of the circle, so the magnetic moment "sees what other frames see".

Now that I'm feeling I'm on the right track, I need to work this whole helix idea back through the rest of the theory.

Aberration:

Electric Field Aberration comes from considering what happens to the source of the observed uniform electric field when it is moving w.r.t. the observer.

From undergraduate physics recall that an infinite plane of charge produces a uniform electric field, and for the plane stationary w.r.t. the observer, the field points perpendicular to the plane.

But a moving charge does not have an angularly uniform field. Due to length contraction, the field lines are concentrated perpendicular to the charge's velocity, with **E** described by Resnick 4-14 below

$$\mathbf{r} = (x, y, z)$$

$$k = \frac{1}{4\pi\varepsilon_\circ} \qquad \gamma = \frac{1}{\sqrt{1 - (v/c)^2}}$$

$$\mathbf{E} = \frac{kq\mathbf{r}}{(x^2 + y^2 + z^2)^{3/2}} \qquad \mathbf{E} = \frac{q\gamma\mathbf{r}}{4\pi\varepsilon_\circ(\gamma^2 x^2 + y^2 + z^2)^{3/2}} \qquad (4\text{-}14)$$

So with components of velocity both parallel and perpendicular to the plane, I make a picture of the plane by rotating the above picture such that the currently vertically oriented flux line is perpendicular to the observer's velocity, then repeat that rotated pattern across the whole horizontal plane.

Velocity of charge $\mathbf{C} = (C_x, C_y, C_z)$

Integrating over the plane will still yield a uniform field, but one with an x-component.

Start with the vector function describing the length contracted field, and rotate it. Then integrate the rotated function across the infinite (x,z)-plane to get the aberrated field.

Rotating a vector function requires rotating both the coefficient functions, and the basis vectors. To do this, I looked up "A first course in general relativity" by Bernard F. Schutz (Not a single undergraduate text on my shelf tells how to rotate a vector function, not even the Vector Calculus text, just how to rotate functions.)

Rotating a vector function requires rotating both the functions in the coefficients and the basis vectors.

First look at **Rotating the functions.**

Rotate a function by θ with the usual matrix formula on the coordinates;

$$\begin{pmatrix} x'' \\ y'' \end{pmatrix} = \begin{pmatrix} \cos\theta & \sin\theta \\ -\sin\theta & \cos\theta \end{pmatrix} \begin{pmatrix} x \\ y \end{pmatrix}$$

or, multiplying from the left by the inverse matrix,

$$\begin{pmatrix} \cos\theta & -\sin\theta \\ \sin\theta & \cos\theta \end{pmatrix}\begin{pmatrix} x'' \\ y'' \end{pmatrix} = \begin{pmatrix} x \\ y \end{pmatrix}$$

Then, multiplying out gives $x = x''\cos\theta - y''\sin\theta$, and $y = x''\cos\theta + y''\sin\theta$ to plug into the x and y of the function.

Looking at the function of a line $y = mx+b$, say horizontal $(m = 0)$ and crossing y-axis at 5 $(b = 5)$;

$y = (0)x+5$, plugging in gets

$(x''\cos\theta + y''\sin\theta) = (0)(x''\cos\theta - y''\sin\theta) + 5$

For simplicity set $\theta = 45°$ so;

$\cos\theta = \sin\theta = \dfrac{\sqrt{2}}{2}$, and,

$$\left(x''\frac{\sqrt{2}}{2} + y''\frac{\sqrt{2}}{2}\right) = (0)\left(x''\frac{\sqrt{2}}{2} - y''\frac{\sqrt{2}}{2}\right) + 5 \ , \ \text{Solving for y'';}$$

$$y'' = -x'' + \frac{10}{\sqrt{2}}$$

Note that the slope of the new line is -1 or -45 degrees where it was 0 or horizontal before (So, a Clockwise rotation).

With the rotation angle being +45 degrees, there are two ways to interpret this

1) The axis rotates +45 degrees
2) The function rotates -45 degrees

Now look at **Rotating the Basis Vectors**. As shown above, a given rotational transformation rotates the coordinates in the functions one way or the basis vectors the other way. Thus I can guess that whatever I transform the functions with, I'll use the inverse to transform the basis vectors. Which is indeed the case shown is Schutz's. Assuming the use of "Einstein's Summation

Notation", where a top and bottom identical idex assumes a sum over that index, but just over the 3 spatial coordinates; Let the rotation matrix to the top left be $R^{x"}{}_x$ and the one to the right be $R^x{}_{x"}$

Schutz 2.4 $x" = R^{x"}{}_x x$

Thus the sum is over the non-primed variables to equal a double primed variable.

Schutz 2.11 $\mathbf{A} = A^\alpha \mathbf{e}_\alpha$

Note that $\mathbf{e}_\alpha$ is a whole basis vector.

Multiplying through by the rotational matrices in the way, as Schutz notes, that "allows the summation notation to work" (So I want the non-primes to cancel), I get

$$R^{x"}{}_x R^x{}_{x"} \mathbf{A} = R^{x"}{}_x A^x R^x{}_{x"} \mathbf{e}_x$$

Noting that the two rotation matrices are inverses of each other,

$$\mathbf{A} = R^{x"}{}_x A^x R^x{}_{x"} \mathbf{e}_x$$

A, as a vector stands independent of any coordinate system. Here it is now expressed in the double primed system.

With the coefficients being non-linear functions of the position vectors (Coordinates), I can't just multiply by the transform out front, but I must substitute with Schutz 2.4. Basically multiplying every instance of the coordinate in the function by it's transform.

Since the basis vectors are linear functions of the coordinates, for them, simply multiply by the inverse matrix.

I wish to rotate the length contracted field from Resnick 4-14 Counter Clockwise, or $+ \theta$, so I need the inverse of the above transform for the coefficient functions, simply swap non-primes for double primes and then substitute into the equation like I did for the line example, and then transform the basis vectors by simply multiplying the result by the inverse transform.

A Tachyon Theory of Everything

Swapping primes, and including the z-dimension, I get;

$$\begin{pmatrix} x \\ y \\ z \end{pmatrix} = \begin{pmatrix} \cos\theta & \sin\theta & 0 \\ -\sin\theta & \cos\theta & 0 \\ 0 & 0 & 1 \end{pmatrix}\begin{pmatrix} x" \\ y" \\ z" \end{pmatrix} \text{ or } \begin{array}{l} x = x"\cos\theta + y"\sin\theta \\ y = -x"\sin\theta + y"\cos\theta \\ z = z" \end{array}$$

Starting with Resnick 4 − 14 for the field of a moving charge;

$$\vec{E} = \frac{q\gamma\vec{r}}{4\pi\varepsilon_o(\gamma^2 x^2 + y^2 + z^2)^{\frac{3}{2}}} \quad . \text{ Then substitute in the above;}$$

$$\vec{E} = \frac{q\gamma(x"\cos\theta + y"\sin\theta, \; -x"\sin\theta + y"\cos\theta, \; z")}{4\pi\varepsilon_o\left(\gamma^2(x"\cos\theta + y"\sin\theta)^2 + (-x"\sin\theta + y"\cos\theta)^2 + z"^2\right)^{\frac{3}{2}}}$$

Using $\gamma = \dfrac{1}{\sqrt{1-\beta^2}}$, $k_\varepsilon = \dfrac{1}{4\pi\varepsilon_o}$,and algebra on the denominator;

$$\vec{E} = \frac{k_\varepsilon q\gamma(x"\cos\theta + y"\sin\theta, \; -x"\sin\theta + y"\cos\theta, \; z")}{\left(\gamma^2(x"^2(1-\beta^2\sin^2\theta) + y"^2(1-\beta^2\cos^2\theta) + 2x"y"\beta^2\cos\theta\sin\theta) + z"^2\right)^{\frac{3}{2}}}$$

Now to complete the rotation by rotating the basis vectors with a multiplication by the inverse matrix. For starters let the whole expression outside the vector parentheses in the numerator be S, then $\mathbf{E} = S(x"\cos\theta + y"\sin\theta, -x"\sin\theta + y"\cos\theta, z")$. Then multiply by the inverse matrix out front to get the rotated $\mathbf{E}$, $\mathbf{E_R}$;

$$\vec{E}_R = \begin{pmatrix} \cos\theta & -\sin\theta & 0 \\ \sin\theta & \cos\theta & 0 \\ 0 & 0 & 1 \end{pmatrix} S(x"\cos\theta + y"\sin\theta, \; -x"\sin\theta + y"\cos\theta, \; z")$$

80

Appendix Magnetic Force Aberration

$$\vec{E}_R = S \begin{pmatrix} \cos\theta & -\sin\theta & 0 \\ \sin\theta & \cos\theta & 0 \\ 0 & 0 & 1 \end{pmatrix} \begin{pmatrix} x''\cos\theta + y''\sin\theta \\ -x''\sin\theta + y''\cos\theta \\ z'' \end{pmatrix}$$

$$\vec{E}_R = S \begin{pmatrix} (\cos\theta)(x''\cos\theta + y''\sin\theta) + (-\sin\theta)(-x''\sin\theta + y''\cos\theta) \\ (\sin\theta)(x''\cos\theta + y''\sin\theta) + (\cos\theta)(-x''\sin\theta + y''\cos\theta) \\ z'' \end{pmatrix} = S \begin{pmatrix} x'' \\ y'' \\ z'' \end{pmatrix}$$

$$\vec{E}_R = \frac{k_\varepsilon q \gamma (x'', y'', z'')}{\left(\gamma^2 \left(x''^2(1 - \beta^2\sin^2\theta) + y''^2(1 - \beta^2\cos^2\theta) + 2x''y''\beta^2\cos\theta\sin\theta \right) + z''^2 \right)^{\frac{3}{2}}}$$

This is the rotated electric field of a moving charge. Now to integrate it over the plane. For convenience, drop the double primes and just call the vector **E**, then displace the equation by (h, j, k) with the substitutions, x = x-h, y = y-j, and z = z-k.

I present the following page sideways to fit the equations on the page legibly;

$$\vec{E} = \frac{k_\varepsilon q \gamma (x - h, y - j, z - k)}{\left(\gamma^2((x-h)^2(1 - \beta^2\sin^2\theta) + (y - j)^2(1 - \beta^2\cos^2\theta) + 2(x - h)(y - j)\beta^2\cos\theta\sin\theta) + (z - k)^2\right)^{\frac{3}{2}}}$$

Plane of charge is in (x,z) - plane, so no y displacement and $j = 0$. Wish to know $\vec{E}$ at some position, y, above the origin, so evaluate at (0, y, 0) setting $x = z = 0$. Replace q with $\sigma dhdk$, σ = area density of plane's charge, then the field being calculated is an element, dE, of the electric field. The rest frame of the plane has a transformed density dealt with later.

$$d\vec{E} = \frac{k_\varepsilon \sigma dhdk \gamma (-h, y, -k)}{\left(\gamma^2((-h)^2(1 - \beta^2\sin^2\theta) + (y)^2(1 - \beta^2\cos^2\theta) + 2(-h)(y)\beta^2\cos\theta\sin\theta) + (-k)^2\right)^{\frac{3}{2}}}$$

Add the integration symbols to sum over the entire (x,z) - plane;

$$\vec{E} = \sigma \gamma k_\varepsilon \int\limits_{-\infty}^{+\infty}\int\limits_{-\infty}^{+\infty} \frac{(-h, y, -k)dhdk}{\left(\gamma^2((-h)^2(1 - \beta^2\sin^2\theta) + (y)^2(1 - \beta^2\cos^2\theta) + 2(-h)(y)\beta^2\cos\theta\sin\theta) + (-k)^2\right)^{\frac{3}{2}}}$$

Appendix Magnetic Force Aberration

Integrating first w.r.t. h using Reference 2 page 83, equations 2.264 5 and 6, but with

$x = h$, and; $R = a + bh + ch^2$ and $\Delta = 4ac - b^2$ where

$a = y^2\gamma^2(1-\beta^2\cos^2\theta) + k^2$, $b = -2\gamma^2 y\beta^2\cos\theta\sin\theta$, and $c = \gamma^2(1-\beta^2\sin^2\theta)$.

Then the x-component will use solution 2.264 6, while the y and z-components will use 2.264 5

Thus for the x-component;

$$\int_{-\infty}^{+\infty} \frac{-h}{\sqrt{R^3}} dh = \left. \frac{--2(2a + bh)}{\Delta\sqrt{R}} \right|_{-\infty}^{+\infty}$$

To take the limits at $\pm\infty$, divide top and bottom by the absolute value of h, $|h|$. The one in the denominator disappears under the square root sign since the term is squared, but up top, I use the fact that for $h < 0$, $h/|h| = -1$ while for $h > 0$, $h/|h| = +1$;

$$\int_{-\infty}^{+\infty} \frac{-h}{\sqrt{R^3}} dh = \left. \frac{(4a + 2bh)\dfrac{1}{|h|}}{\Delta\sqrt{a + bh + ch^2}\dfrac{1}{|h|}} \right|_{-\infty}^{+\infty}$$

$$\int_{-\infty}^{+\infty} \frac{-h}{\sqrt{R^3}} dh = \lim_{h\to\infty} \frac{\dfrac{4a}{|h|} + 2b}{\Delta\sqrt{\dfrac{a}{h^2} + \dfrac{b}{h} + c}} - \lim_{h\to-\infty} \frac{\dfrac{4a}{|h|} - 2b}{\Delta\sqrt{\dfrac{a}{h^2} + \dfrac{b}{h} + c}} = \frac{4b}{\Delta\sqrt{c}}$$

Then for the y and z - components;

$$\int_{-\infty}^{+\infty} \frac{dh}{\sqrt{R^3}} = \left. \frac{2(2ch + b)}{\Delta\sqrt{R}} \right|_{-\infty}^{+\infty} = \lim_{h\to\infty} \frac{4c + \dfrac{2b}{|h|}}{\Delta\sqrt{\dfrac{a}{h^2} + \dfrac{b}{h} + c}} - \lim_{h\to-\infty} \frac{-4c + \dfrac{2b}{|h|}}{\Delta\sqrt{\dfrac{a}{h^2} + \dfrac{b}{h} + c}}$$

$$\int_{-\infty}^{+\infty} \frac{dh}{\sqrt{R^3}} = \frac{8\sqrt{c}}{\Delta}$$

Simplifying $\Delta = 4ac-b^2$

$\Delta = 4(y^2\gamma^2(1-\beta^2\cos^2\theta) + k^2)(\gamma^2(1-\beta^2\sin^2\theta))-(-2\gamma^2 y\beta^2\cos\theta\sin\theta)^2$

$\Delta = 4\gamma^2[y^2 + k^2(1-\beta^2\sin^2\theta)]$

Plug everything back into equation and re-arrange for integration w.r.t. k;

$$\vec{E} = \frac{\sigma\gamma}{4\pi\varepsilon_0}\int_{-\infty}^{+\infty}\left(\frac{4b}{\Delta\sqrt{c}}, \frac{y8\sqrt{c}}{\Delta}, \frac{-k8\sqrt{c}}{\Delta}\right)dk$$

Where $\Delta = 4\gamma^2\left(y^2 + k^2(1-\beta^2\sin^2\theta)\right)$,
$b = -\gamma^2 y2\beta^2\cos\theta\sin\theta$, and $c = \gamma^2(1-\beta^2\sin^2\theta)$

$$\vec{E} = \frac{\sigma\gamma}{4\pi\varepsilon_0}\left(\int_{-\infty}^{+\infty}\frac{a_1 ydk}{y^2 + bk^2}, \int_{-\infty}^{+\infty}\frac{a_2 ydk}{y^2 + bk^2}, \int_{-\infty}^{+\infty}\frac{a_2 kdk}{y^2 + bk^2}\right) \text{ where}$$

$$a_1 = \frac{-2\beta^2\cos\theta\sin\theta}{\gamma\sqrt{1-\beta^2\sin^2\theta}} , \quad a_2 = \frac{2}{\gamma}\sqrt{1-\beta^2\sin^2\theta} \text{ and } b = 1-\beta^2\sin^2\theta$$

For the x and y − components, I'm using my old Calculus book;

$$\int_{-\infty}^{+\infty}\frac{dk}{y^2 + bk^2} = \frac{1}{\sqrt{y^2 b}}Arctan\left(k\sqrt{\frac{b}{y^2}}\right)\Bigg|_{-\infty}^{+\infty} = \frac{\pi}{y\sqrt{b}}$$

Continuing with the z-component (using u substitution);

$$\int_{-\infty}^{+\infty}\frac{kdk}{y^2 + bk^2} = \frac{1}{2b}\ln(y^2 + bk^2)\Bigg|_{-\infty}^{+\infty} = 0$$

Putting all the components together, renaming the vector **E** to be **E**Aberrated, and simplifying;

Appendix Magnetic Force Aberration

$$\vec{E}_{\text{Aberrated}} = \frac{\sigma}{2\varepsilon_0}\left(\frac{-\beta^2\cos\theta\sin\theta}{(1-\beta^2\sin^2\theta)},\ 1,\ 0\right)$$

I will call this the "Aberrated E Field" formula
Notice no coordinates left. It's a uniform constant field but with a negative x-component.

Applying the above aberrated field requires some interpretation of a couple of its terms. For the $\beta\cos\theta$, I will transform to the rest frame of the velocity difference (described in the next section). To get the invariant velocity difference between the flux and charge (Shown in next section), then transform that velocity difference back to observer's frame. And for the $\beta\sin\theta$ I will use the "catch up" velocity (described in Velocity appendix).

Thus for Case 5, I will catch up to the velocity difference to give each one (The charge and the flux) the term $\gamma_v C_x$ and $\gamma_v f_x$ where v is the "catch up" velocity mentioned later then take their "Invariant Velocity Difference (explained in Velocity appendix);

$$\frac{\gamma_v C_x - \gamma_v f_x}{1 - \gamma_v C_x \gamma_v f_x}$$

Then I transform this "velocity" back to "my" frame by dividing by the γ_v since I'm now "running away" from the velocity difference, resulting in;

$$\frac{C_x - f_x}{1 - \gamma_v C_x \gamma_v f_x} = \beta\cos\theta$$

Note that when speaking of the velocity of a flux line, there is no notion of moving parallel to it, indeed, put yourself in a static electric field (No **B** field) generated by an infinite plane of charge perpendicular to it, and right under your feet. Now boost to a new velocity that is parallel to the electric field and see that the transformed **E** and **B** fields are identical to those of the stationary frame, so you can again assume an infinite plane of charge generating the electric field right under your feet.

Once there is a **B** field, then there is lateral movement of the infinite plane and it's no longer always under foot since that lateral movement will get transformed with a parallel to flux line boost.

It will be convenient to define the flux's perpendicular velocity vector as $\mathbf{f_o}$ and then define the flux's velocity vector, $\mathbf{f} = \mathbf{f_o} + (\mathbf{C}\cdot\overline{\mathbf{E}})\overline{\mathbf{E}}$ where $(\mathbf{C}\cdot\overline{\mathbf{E}})\overline{\mathbf{E}}$ ($\overline{\mathbf{E}}$ being the unit vector in the direction of the electric field.) is the component of the charge's velocity that is parallel to the flux lines. Basically, since there is no perception of moving parallel to a line, I assume the charge only sees the perpendicular movement of the line and I assume that the flux line's velocity parallel to itself is equal to that of the charge. Thus "Catching Up" to the invariant velocity difference will involve a boost parallel to the flux line that is equal to the charge's component parallel to the flux line.

In the aberration formula, $\beta\sin\theta$ becomes my observed value for the charge's component parallel to the flux lines, the $(\mathbf{C}\cdot\overline{\mathbf{E}})\,\overline{\mathbf{E}}$ Shown above.

To orient the picture, I note that; Given an arbitrary $\mathbf{f_o}$ and $\mathbf{C}$ with $\mathbf{f_o}$ perpendicular to $\mathbf{E}$, The vector $(\mathbf{f} - \mathbf{C})$ is perpendicular to $\mathbf{E}$ since $(\mathbf{f} - \mathbf{C})\cdot\overline{\mathbf{E}} = (\mathbf{f_o} + (\mathbf{C}\cdot\overline{\mathbf{E}})\overline{\mathbf{E}})\cdot\overline{\mathbf{E}} - \mathbf{C}\cdot\overline{\mathbf{E}}$
$= \mathbf{f_o}\cdot\overline{\mathbf{E}} + (\mathbf{C}\cdot\overline{\mathbf{E}})\overline{\mathbf{E}}\cdot\overline{\mathbf{E}} - \mathbf{C}\cdot\overline{\mathbf{E}} = 0$ since $\mathbf{fo}\cdot\overline{\mathbf{E}} = 0$ by definition of $\mathbf{f_o}$, and $\overline{\mathbf{E}}\cdot\overline{\mathbf{E}} = 1$. Thus I can rotate my general case so that $\mathbf{f} - \mathbf{C}$ lies along the x-axis and $\mathbf{E}$ points in the y-direction. Then $\mathbf{f} - \mathbf{C} = f_x - C_x$ with $(\mathbf{C}\cdot\overline{\mathbf{E}})\overline{\mathbf{E}} = C_y = f_y$, and $f_z = C_z$.

Appendix Magnetic Force Pressure Factor

Pressure Factor P_f

The $\sigma/2\varepsilon_o$ out in front of the aberrated vector is from the rest frame of the plane and is set equal to E_o, the static electric field in the rest frame of the plane. But if the charge sees moving flux lines, then the charge sees the flux field get length contracted and thus concentrated by γ_v where v is the velocity the charge sees the flux moving at. Since I am making a pressure times area model, and I call the "**rest pressure**", $P_o = E_o/\pi a^2$, and I call the above γ_v term the "**Pressure Factor**", P_f.

I want γ_v where v is the velocity difference between the flux and the charge. For some frames the geometry is simple to trivial, but for other frames it proves more convenient to transform the pressure factor from a simpler frame to the frame of interest. So I use the transforms on a simple case to derive the pressure factor for the most general case described above to have a term applicable in all frames.

Starting with the pressure factor from Case 4 for the s-frame where $P_f = \gamma_c/\gamma_y$. Then boost the s'- frame in the x-direction with an arbitrary velocity $v = -f_{x'}$ since f_x is standing still in the s-frame.

$$P_f = \sqrt{\frac{1 - C_y^2}{1 - C_x^2 - C_y^2}} \quad , \quad C_x = \frac{C_{x'} + v}{1 + C_{x'}v} \quad , \quad C_y = \frac{C_{y'}\sqrt{1 - v^2}}{1 + C_{x'}v} \quad , \quad v = -f_{x'}$$

$$P_{f'} = \sqrt{\frac{1 - \dfrac{C_{y'}^2(1 - v^2)}{(1 + C_{x'}(-f_{x'}))^2}}{1 - \left(\dfrac{C_{x'} + (-f_{x'})}{1 - C_{x'}f_{x'}}\right)^2 - \dfrac{C_{y'}^2(1 - f_{x'}^2)}{(1 - C_{x'}f_{x'})^2}}} = \sqrt{\frac{(1 - C_{x'}f_{x'})^2 - C_{y'}^2(1 - f_{x'}^2)}{(1 - C_{x'}^2 - C_{y'}^2)(1 - f_{x'}^2)}}$$

Then boost with arbitrary $v' = -C_{z''} = -f_{z''}$, so;

$$C_{x'} = \frac{C_{x''}\sqrt{1-v'^2}}{1+C_{z''}v'} \quad , \quad C_{y'} = \frac{C_{y''}\sqrt{1-v'^2}}{1+C_{z''}v'} \quad , \quad f_{x'} = \frac{f_{x''}\sqrt{1-v'^2}}{1+f_{z''}v'}$$

Plug these into $P_{f''}$ including subbing $C_{z''}$ for both $-v'$ and $f_{z''}$ but drop the double primes since all are double prime;

$$P_f = \sqrt{\frac{\left[1-\left|\frac{f_x\sqrt{1-C_z^2}}{1+C_z(-C_z)}\right|\left(\frac{C_x\sqrt{1-C_z^2}}{1-C_z^2}\right)\right]^2 - \left|\frac{C_y\sqrt{1-C_z^2}}{1-C_z^2}\right|^2\left[1-\left(\frac{f_x\sqrt{1-C_z^2}}{1-C_z^2}\right)^2\right]}{\left[1-\left(\frac{C_x\sqrt{1-C_z^2}}{1-C_z^2}\right)^2 - \left|\frac{C_y\sqrt{1-C_z^2}}{1-C_z^2}\right|^2\right]\left[1-\left(\frac{f_x\sqrt{1-C_z^2}}{1-C_z^2}\right)^2\right]}}$$

After some algebra;

$$P_f = \sqrt{\frac{1-2C_z^2-2f_xC_x+C_z^4+2f_xC_xC_z^2+f_x^2C_x^2-C_y^2+C_y^2C_z^2+f_x^2C_y^2}{\left(1-(C_x^2+C_y^2+C_z^2)\right)\left(1-(f_x^2+f_z^2)\right)}}$$

Looking at the bottom terms, $C_x{}^2 + C_y{}^2 + C_z{}^2 = \mathbf{C \cdot C}$, since $\mathbf{E}$ points in y-direction, $f_x{}^2 + f_z{}^2 = \mathbf{f_o \cdot f_o} = \mathbf{f \cdot f} - (\mathbf{C \cdot E})^2$. Then, adding $+ C_x{}^2 C_y{}^2 - C_x{}^2 C_y{}^2$ to the top, and using the facts that most of the top terms can be made to equal $(1 - \mathbf{f \cdot C})^2$ and that in this orientation, $C_y = \mathbf{C \cdot \overline{E}}$, and $(f_x - C_x) = \mathbf{f} - \mathbf{C}$ and $(\mathbf{f} - \mathbf{C}) \cdot (\mathbf{f} - \mathbf{C}) = (f_x - C_x)^2$, then with more algebra I get;

$$P_f = \sqrt{\frac{\left(1-\vec{f}\cdot\vec{C}\right)^2 + \left(\vec{f}-\vec{C}\right)\cdot\left(\vec{f}-\vec{C}\right)\left(\vec{C}\cdot\overline{E}\right)^2 + \left(1-\vec{C}\cdot\vec{C}\right)\left(\vec{C}\cdot\overline{E}\right)^2}{\left(1-\vec{C}\cdot\vec{C}\right)\left(1-\vec{f}\cdot\vec{f}+\left(\vec{C}\cdot\overline{E}\right)^2\right)}}$$

Making the pressure factor a combination of dot products of vectors from the frame of interest.

Velocity; A discussion and a short cut.
Velocity Difference

Starting with the velocity addition equation;

$$U_x = \frac{U_x' + v}{1 + \dfrac{U_x' v}{c^2}}$$

The U_x is the addition of the s'-frame's velocity, v, to the s'-frame's particle's velocity $U_{x'}$, thus the inverse transform;

$$U_x' = \frac{U_x - v}{1 - \dfrac{U_x v}{c^2}}$$

is the difference between the s-frame's particle's velocity, U_x, and the s'-frame's speed, v.

So call $U_{x'}/c$ the velocity difference between the two particles, one at speed $U_x/c = a$, and the other at speed $v/c = b$;

$$\frac{a - b}{1 - ab}$$

Where setting either to zero leaves just the other with a one in the denominator. But take this form and transform it to any other frame moving at speed, v, along the x-axis;

$$\frac{a' - b'}{1 - a'b'} = \frac{\left(\dfrac{a-v}{1-av}\right) - \left(\dfrac{b-v}{1-bv}\right)}{1 - \left(\dfrac{a-v}{1-av}\right)\left(\dfrac{b-v}{1-bv}\right)} = \frac{a - b}{1 - ab}$$

and I get back what I started with. I will call this the "**Invariant velocity difference**". As long as the line defined by the velocity difference is not moving perpendicular to itself, then I can claim to be in the "rest frame of the velocity difference", and I can take my observed values and plug them into the equation to get the invariant velocity difference.

For two objects with arbitrary velocities, it is necessary to first transform to the **"Rest frame of the velocity Difference"**. Rotate the frame so the velocity vectors are in the (x,y)-plane and their difference is along the x-axis. Then **"catch up"** to the "rest frame of the velocity difference" with a boost velocity of C_y.

In the following diagram, the U_y can be said to be moving away from the observer at speed U_x. By boosting to the s'-frame with velocity $v = U_x$, I am **"catching up"** to the U_y velocity, as seen in the calculation, this **"caught up"** velocity is $U_{y'} = \gamma U_y$.

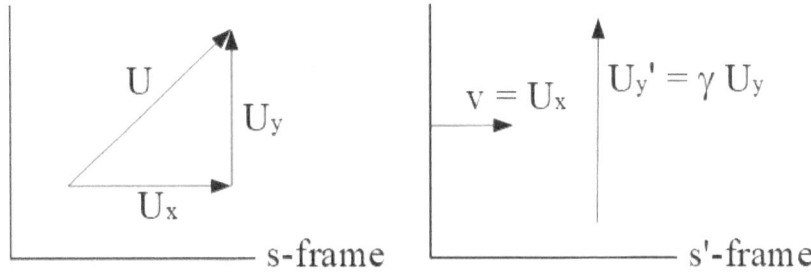

Then, with all velocities as β's;

$$U_{y'} = \frac{U_y\sqrt{1-v^2}}{1-U_x v} = \frac{U_y\sqrt{1-U_x^2}}{1-U_x U_x} = \gamma_{U_x} U_y$$

Symmetrically, the same can be said about U_x moving away from the observer at speed U_y with $U_{x''} = \gamma_{U_y} U_x$, and s"-frame having boost velocity $v' = U_y$ in the y-direction. Thus I will use the term **"catch up"** to describe transforming to the frame where that is the only component.

So, I will use the following short cut;
"Catch Up" to a velocity will **multiply** by the appropriate γ, and **"Run Away"** from a velocity will the **divide** by the appropriate γ.

Relativistic Mass

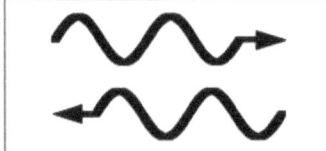 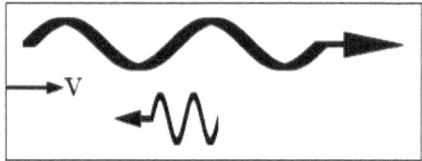

S-frame is stationary   S'-frame is moving to the right at speed v

Two photons of equal energy   Top photon is red shifted.
going in opposite directions.   Bottom photon is blue shifted.

Top photon has energy, $E_1 = h\nu_1$ and bottom photon, $E_2 = h\nu_2$ with $\nu_1 = \nu_2 = \nu$ in the s-frame and the total rest mass, m_o, of the center of mass of the mass less box, is

$$E/c^2 = (E_1 + E_2)/c^2 = (h\nu + h\nu)/c^2 = 2h\nu/c^2 .$$ Then the top photon in the s'-frame has frequency

$$\nu_{1'} = \nu_1 \sqrt{\frac{(1-\beta)}{(1+\beta)}} \ , \ \text{while for the bottom, } \nu_{2'} = \nu_2 \sqrt{\frac{(1+\beta)}{(1-\beta)}}$$

with $\beta = \dfrac{v}{c}$. Summing these mass/energies gives;

$$\frac{(E_{1'} + E_{2'})}{c^2} = \frac{(h\nu_{1'} + h\nu_{2'})}{c^2} = \frac{h(\nu_{1'} + \nu_{2'})}{c^2} \ , \ \text{then substitute;}$$

$$\frac{h}{c^2}\left[\nu_1\sqrt{\frac{(1-\beta)}{(1+\beta)}} + \nu_2\sqrt{\frac{(1+\beta)}{(1-\beta)}}\right] = \frac{h\nu}{c^2}\left(\frac{2}{\sqrt{1-\beta^2}}\right) = \frac{E\gamma}{c^2} = \gamma m_o$$

So, $m' = \gamma m_o$ just like a relativistic mass.

Then summing the momenta will show the **Inertial mass**

by starting with the usual $E = pc$ or $p = E/c$ then in the s'-fame $p_1' = h\nu'_1/c$ and $p_2' = -h\nu'_2/c$ .Summing gives $p' = p_1' + p_2' = h\nu'_1/c - h\nu'_2/c$. Then using the same substitutions as above:

$$p' = \frac{h}{c}\nu_1\sqrt{\frac{(1-\beta)}{1+\beta}} - \frac{h}{c}\nu_2\sqrt{\frac{(1+\beta)}{1-\beta}} = -\frac{h}{c}\frac{2\nu\beta}{\sqrt{1-\beta^2}}\frac{c}{c} = -\frac{E}{c^2}\beta c\gamma$$

$= -\gamma m_o\beta c$ again, just like a relativistic momentum with inertial mass γm_o. The negative comes from the fact that when I get on my s'-frame moving to the right (positive direction) at speed, v, I see the box moving to the left (negative direction) at speed, v.

Gravitational Mass

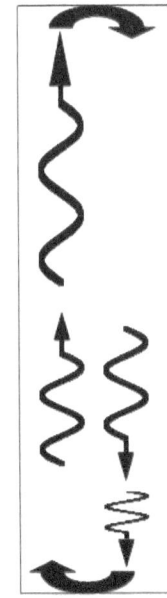

Red shifted photon at the top of the box

A photon of frequency ν does one round trip from the middle of the box,
to the top of the box,
then reflect down to the bottom of the box,
reflecting back up to the middle again.

Blue shifted photon at the bottom of the box

To get the force of gravity on the box, I'm going to use

92

Appendix Photons in a Box

$F = \Delta p/\Delta t$, where Δp will be the net momentum transferred to the box in one round trip of the photon and Δt will be the time of that round trip. Using Resnick's equation (C-2), $v' = v \pm \Delta v$, where $\Delta v = v\gamma d/c^2$. Then $\Delta p = p_{top} - p_{bottom}$, and $p = 2hv/c$ (The 2 is for the reflection.). Since I'm starting the round trip in the middle of the box, the d in his equation $= L/2$ below. Also the subscript, t, is for the top of the box while the subscript, b, is for the bottom of the box.

$$\Delta p = p_t - p_b = \frac{2hv_t}{c} - \frac{2hv_b}{c} = \frac{2h}{c}\left(v\left(1 - \frac{gL}{2c^2}\right) - v\left(1 + \frac{gL}{2c^2}\right)\right)$$

$$\Delta p = \frac{-2hvgL}{c^3} \text{ ,and } \Delta t = \frac{2L}{c} \text{ , so}$$

$$F = \frac{\dfrac{-2hvgL}{c^3}}{\dfrac{2L}{c}} = \frac{-hvg}{c^2} = \frac{-Eg}{c^2} = -mg$$

like a mass in a gravity field.

   Again, the calculations are nothing new. My model simply reverses the assumption from looking at how photons in a box behave like a mass to looking at the mass as a photon in a box. Or at least a photon bouncing around due to the path of it's tachyon. I once heard that Einstein thought that matter might be knots in time, If he meant "knots in space-time" (which, being the one who tied space and time together he probably did), then, with tachyons being elements of space time, he was presciently correct.

A Tachyon Theory of Everything

My thoughts on Shape began with Self Interference and then, with the v = (c, U) notation, I began to explore particles and concluded that neutral particles (except neutrinos) shared opposing tachyons that could drift apart, and that neutrinos were available states that had no tachyon in them.
Continuing with other notions from the theory;
If I got the pressure of the cloud causing the tachyons to turn and twist out of each other's way, then I also need to make sure that the pressure modulations due to the gravity (tachyon mass energy density gradients) gradients it will fly through on the way around the galaxy/universe does not alter the path enough to miss it's return point. One possibility is if pressure modulates the rest mass, but not the speed or arc, another possibility is if the loops are so big and the cloud so thick that even a galaxy is a too mild an influence on the path. Anyway, putting that concern to the side for now, I figure;
The tachyon gets pushed into a curve by the cloud's pressure, (Another possibility, restricting the paths to circles, is not bending the path, but only imparting an angular momentum at the point of self interference.)
The tachyons different sides may get different curvatures
That photons get smaller with larger energy makes me think the curvatures are proportional (and maybe to some of it's sides inversely proportional or some functional relation between the sides) to the mass/energy of the tachyon. Thus higher masses will have smaller radii of curvature.
From my Differential Geometry (Kreyszig) I see that the only shapes I can get from a uniform cloud pressure and thus uniform curvature is circles and circular helices.
Circles may be the model, with rationalizations that ignore the particle's single tachyon's contribution to it's own field (reasonable), and that photons have tachyon loops to one side, and

that the un-canceled momentum is undetectable (hard to do).

Besides, I've had my models for decades in witch the legs of my photon and electron models have the return leg with the opposite velocity which rules out circles,

So to get more complex shapes, I rationalize a **Self Pressure** acting in addition to the cloud's pressure. Since the self interfering particle is always there, there is a pressure gradient due to that particle's own tachyon path, so the curvature can be a function of the distance from the self interfered point. The cardioid, with a curvature equal to the inverse square root of the distance to the cusp can meet such a requirement.(I have yet to explore what other functional relations of curvature $\sim r^n$ where r is the distance from the self interfered particle to the tachyon path point and n is an arbitrary number.). And although the derivatives of the flipped cardioids match at the cusps, I don't like the infinite curvature at the point of self interference, since I want the curve to locally look straight.

Never-the-less, the illustration (along with some bent wire) is useful for a possible mental image. Two cardioids flipped w.r.t each other and joined at the cusps so that the tachyon, instead of changing direction at the cusp, continues through to the other cardioid.

The arrows show a net counterclockwise direction so repeating the pattern for the small helix action still allows for the electric field.

Since any plane closed curve will have an arc length that is linearly proportional to a characteristic radius, I will just assume whatever the shape, it will have a characteristic radius, r, distinguishing the relative size of similar shapes. So in further discussion I will simply assume everything is circles.

    Using the photon as a model (See below), I figure the small helix is the **E** field for half the photon. So a photon , made up of 2 tachyons, has energy E_p = hc/λ = Plank's constant times the speed of light divided by the wavelength of the photon. but the whole photon is two passes of the tachyon so as a first guess I'll make half the wavelength equal to the diameter of the helix (I'd like to make the helix much smaller, but this guess comes from a physical reality and allows me to work with something, I'm looking for the car keys under the street lamps.). Then, in that helix's rest frame where the path is a plane circle, I'll say that the energy of the tachyon in that rest frame (I'll call the s' frame) has the radius of the helix in that rest frame which is also the same in our frame (not primed), since distance perpendicular to velocity is invariant,

    | Indicates a positive tachyon

    | Indicates a negative tachyon

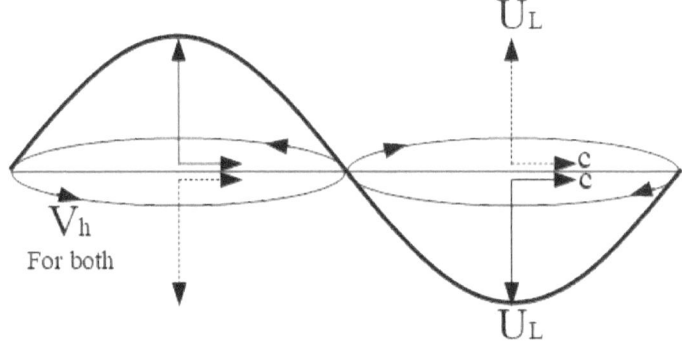

$r_h = r_{h'} = \lambda/4$ so $E_p = hc/4r_h$

    For what follows; γ = the usual special relativity gamma

Appendix Shape

function. The subscript, L, will refer to the tachyon's loop velocity which is so nearly constant ($V_L \ggggg c$ and any changes to it are $\ll c$) that I will treat it as such. The subscript, h, refers to the tachyons helix velocity.

Also $U^2 = V^2 - c^2$ so $\gamma = c/U$ and I'll be ignoring any complex (square root of negative one) notation for now, since I'm mostly concerned with magnitudes.

The energy of a tachyon is, $E_T = \gamma_L \gamma_{h'} m_o c^2$ (m_o = rest mass) and in the rest frame of the helix is, $E_T' = \gamma_{h'} m_o c^2$ where the tachyon is moving in a tiny circle with speed $v_{h'}$. Since there are two tachyons, the photons energy is split between them thus $E_p = 2E_T$ and $hc/4r_h = 2\gamma_L \gamma_{h'} m_o c^2$. The photon's energy transformed to the helix's rest frame is $E_p' = E_p/\gamma_L$, and in this s'-frame, I equate the photon and tachyon's energies; $E_p' = 2E_T'$ or $hc/4r_h \gamma_L = 2\gamma_{h'} m_o c^2$. I'm looking at this equation as containing the energy of a tachyonic particle (to the right of the equal sign, let $\gamma_{h'} m_o c^2$ = (the energy of the tachyonic particle = E), and its radius in the the cloud the r_h, (let $r_h = r$.).

Now $hc/8\gamma_L \gamma_{h'} m_o c^2 = r_h = r = hc/8\gamma_L E$ If this is my formula for the radius of a tachyonic particle with a given energy, then the tachyonic particles making up the photon have half the photon's energy of $hc/2\lambda = hc/2(4rh)$ and plugging in for the E above will give the loop radius of

$r = r_L = hc/(8\gamma_L E) = hc/(8\gamma_L(hc/8r_h)) = r_h/\gamma_L = r_h U_L/c$

the loop radius is U_L times bigger than the helix radius allowing for the same rule to give me photonic sized helix circles and galactic and larger sized loops.

A Tachyon Theory of Everything

Now I assume this planar curve of a slinky the size of a
galaxy or larger moves perpendicular to the page at the speed of
light, c. Although the small helical action is left or right handed
depending on if it's a plus or minus signed tachyon, I'm not sure I
require the same handedness of this larger loop motion's helical
action. This restriction of the perpendicular velocity of the large
loop to ±c may be too simple, but it allows for simpler thinking
and math for now. This makes for the a natural split of the loops
velocity between a U_L going in a giant plane closed path and a c
moving the path along as a helix.

Then I want to bounce or rotate the photon's plane pattern
that is sweeping perpendicular to the page to make a helix, to a
particle pattern that is either bouncing behind and in front of the
page or rotating like the page is spinning. This is where I hope to
find restrictions that will explain the three generations of particles
if not the masses $(V_{h'})$ of the particles themselves.

I think the leptons (sans neutrinos which I think are empty
available states of their more massive lepton counterparts.) are
rotated by $\pi/2$ while the baryons are rotated by $\pi/3$ or $2\pi/3$ (or
some $n\pi$ addition). In the picture below I have bent a wire into
what might be a proton (3 coils of a spring bent around tip to tail
and joining the two ends.) with each of the loops making a quark,
and I imagine the pattern further spinning and precessing.

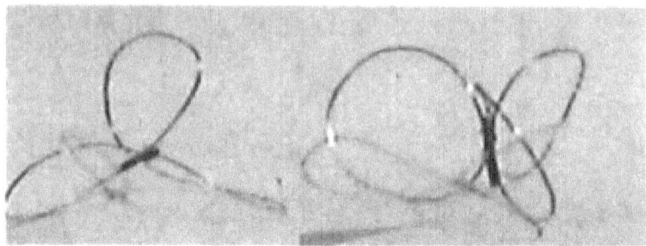

Appendix Shape

That all tachyons have the same loop velocity suggests a heaviest particle where the helix velocity is zero $(m = \gamma_{vL}m_o)$ or 1c (infinite mass?) or 2c.

Then there is the possibility that the helix velocity is less than c.

Another role for Self Pressure might be that it has to be high enough that a tachyon can reenter it's own light cone to self interfere to create a real mass.

But in the several months I've been working on this(I keep getting infinite numbers of solutions for particles), I've realized this will be another 2 or 3 year endeavor (like every other major subject in this theory). so I'm cinching things up where they stand to publish what I've got so far.

TRADITIONAL ANSWERS. For the first three cases I use Resnick's v/c notation

Case 1 In s-frame, $\mathbf{E} = (0, E_o, 0)$, and $\mathbf{B} = (0, 0, 0)$. The charge's velocity is $\mathbf{C} = (0, 0, 0) = (C_x, C_y, C_z)$, thus the velocity, $\mathbf{U}$, in Resnick's formulas is, $U_x = C_x = 0$, so $C_{x'} = (0-v)/1 = -v$, and $U_y = C_y = 0$, so $C_{y'} = 0$, and $U_z = C_z = 0$, so $C_{z'} = 0$, So $\mathbf{C}' = (-v, 0, 0)$.

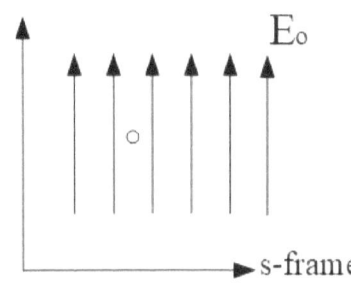
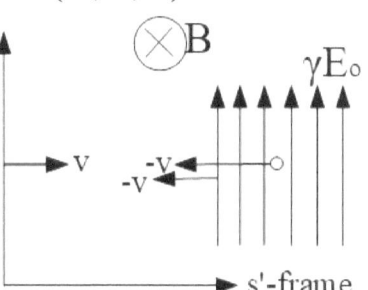

Static charge in a static **E** field   Both the charge and the flux have a velocity of -v

Start with the Force transformation;

$$F_{x'} = \frac{F_x - \frac{v}{c}\left(\vec{C} \cdot \vec{F}\right)}{1 - \frac{C_x v}{c^2}} = \frac{0 - \frac{v}{c}(0 + 0 + 0)}{1 - \frac{0v}{c^2}} = 0$$

$$F_{y'} = \frac{F_y\sqrt{1 - \frac{v^2}{c^2}}}{1 - \frac{C_x v}{c^2}} = \frac{qE_o\sqrt{1 - \frac{v^2}{c^2}}}{1 - 0} = \frac{qE_o}{\gamma} \text{ ,and } F_{z'} = \frac{F_z\sqrt{1 - \frac{v^2}{c^2}}}{1 - \frac{C_x v}{c^2}} = 0$$

$$\vec{F}' = \left(0, \frac{qE_o}{\gamma}, 0\right)$$

Appendix Traditional Answers

Now transform **E** and **B** fields;

$$E_{x'} = E_x \qquad\qquad B_{x'} = B_x$$

$$E_{y'} = \gamma(E_y - vB_z) = \gamma E_o \quad B_{y'} = \gamma\left(B_y + \frac{v}{c^2}E_z\right) = 0$$

$$E_{z'} = \gamma(E_z + vB_y) = 0 \quad B_{z'} = \gamma\left(B_z - \frac{v}{c^2}E_y\right) = \gamma\left(0 - \frac{v}{c^2}E_o\right) = -\gamma\frac{v}{c^2}E_o$$

$$\vec{C'} \times \vec{B'} = \begin{vmatrix} i & j & k \\ -v & 0 & 0 \\ 0 & 0 & -\gamma\frac{v}{c^2}E_o \end{vmatrix} = i(0-0) - j\left(\gamma\frac{v^2}{c^2}E_o - 0\right) + k(0-0)$$

$$\vec{C'} \times \vec{B'} = \left(0, \; -\gamma\frac{v^2}{c^2}E_o, \; 0\right)$$

$$\vec{F'} = q\left(\vec{E'} + \vec{v'} \times \vec{B'}\right) = q\left((0, \gamma E_o, 0) + \left(0, \; -\gamma\frac{v^2}{c^2}E_o, 0\right)\right)$$

$$\vec{F'} = qE_o\left(0, \gamma - \gamma\frac{v^2}{c^2}, 0\right) = qE_o\left(0, \gamma\left(1 - \frac{v^2}{c^2}\right), 0\right) = \left(0, \frac{qE_o}{\gamma}, 0\right)$$

Case 2 has the charge in the s-frame moving up parallel to the **E** field. Same **E** and **B** as Case 1;

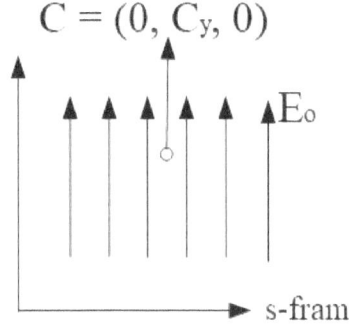

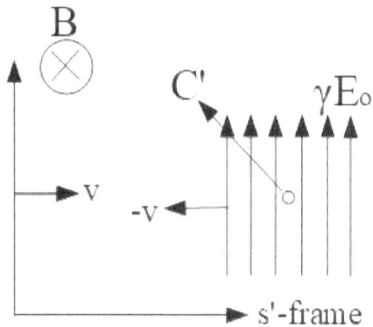

Charge moving parallel
to a static **E** field

Both the charge and the flux
have same x'-velocity of -v

Again, start with transforming the force directly;

$$F_{x'} = \frac{F_x - \frac{v}{c^2}\left(\vec{C} \cdot \vec{F}\right)}{1 - \frac{C_x v}{c^2}} = \frac{0 - \frac{v}{c^2}(0 + C_y qE_0 + 0)}{1 - 0} = \frac{-vC_y qE_0}{c^2}$$

$$F_{y'} = \frac{F_y\sqrt{1 - \frac{v^2}{c^2}}}{1 - \frac{C_x v}{c^2}} = \frac{qE_0\sqrt{1 - \frac{v^2}{c^2}}}{1} = \frac{qE_0}{\gamma} \quad , \quad F_{z'} = 0$$

$$\vec{F'} = \left(\frac{-qE_0 vC_y}{c^2}, \frac{qE_0}{\gamma}, 0\right)$$

For the EM solution, use the same **E, E', B** and **B'** as Case 1.
Transform **C** to **C '** for the Lorentz force;

Appendix Traditional Answers

$$C_{x'} = \frac{C_x - v}{1 - \frac{C_x v}{c^2}} = \frac{0 - v}{1 - 0} = -v \ , \quad C_{y'} = \frac{C_y \sqrt{1 - \frac{v^2}{c^2}}}{1 - 0} = \frac{C_y}{\gamma} \ , \quad C_{z'} = 0$$

$$\vec{C'} \times \vec{B'} = \begin{vmatrix} i & j & k \\ -v & \dfrac{C_y}{\gamma} & 0 \\ 0 & 0 & \dfrac{-v\gamma E_o}{c^2} \end{vmatrix}$$

$$\vec{C'} \times \vec{B'} = i\left(\frac{C_y}{\gamma}\left(\frac{-v\gamma E_o}{c^2}\right) - 0\right) - j\left(\frac{v^2\gamma E_o}{c^2} - 0\right) + k(0 - 0)$$

Plugging into **v'** x **B'** of the Lorentz force

$$\vec{F'} = q\left(\vec{E'} + \vec{v'} \times \vec{B'}\right) = q\left((0, \gamma E_o, 0) + \left(\frac{-vC_y E_o}{c^2}, \frac{-v^2\gamma E_o}{c^2}, 0\right)\right)$$

$$\vec{F'} = q\left(\frac{-vC_y E_o}{c^2}, \gamma E_o - \frac{-v^2\gamma E_o}{c^2}, 0\right) = \left(\frac{-qE_o vC_y}{c^2}, \frac{qE_o}{\gamma}, 0\right)$$

Case 3 has the charge moving horizontally at velocity, v, in the s-frame. So $U_x = C_x = v$

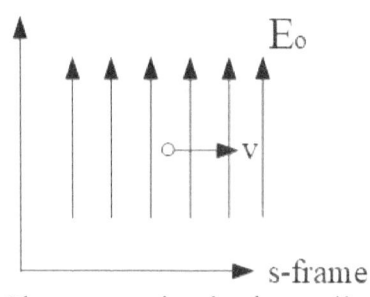 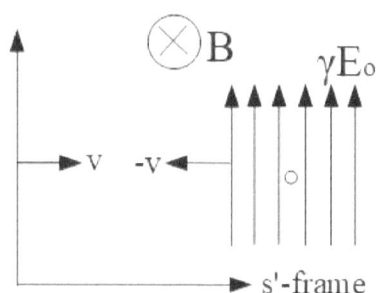

Charge moving horizontally   Charge is stationary while
in a static **E** field   the flux has a velocity of -v

Since $F_x = 0$, and $\mathbf{F} \cdot \mathbf{C} = (F_x C_x + F_y C_y + F_z C_z) = 0v + qE_o 0 + 00 = 0$, so $F_{x'} = 0$, then for $F_{y'}$;

$$F_{y'} = \frac{F_y \sqrt{1 - \frac{v^2}{c^2}}}{1 - \frac{C_x v}{c^2}} = \frac{F_y \sqrt{1 - \frac{v^2}{c^2}}}{1 - \frac{vv}{c^2}} = \gamma F_y$$

$$\mathbf{F'} = (0, \gamma q E_o, 0)$$

Since the charge's velocity in the s' frame is zero,
$\mathbf{C' \times B'} = \mathbf{0 \times B'} = \mathbf{0}$, so $\mathbf{F'} = q(\mathbf{E' + v' \times B'}) = q\mathbf{E'}$

$E_{x'} = E_x = 0$, $E_{y'} = \gamma(E_y - vB_z) = \gamma(E_o - 0) = \gamma E_o$,
$E_{z'} = \gamma(E_z + vB_y) = \gamma(0 + 0) = 0$,

$\mathbf{E'} = (0, \gamma E_o, 0)$, and $\mathbf{F'} = q\mathbf{E'} = (0, q\gamma E_o, 0)$.

From now on all velocities are β's or divided by c.

Case 4 s-frame: The charge going up at an angle.

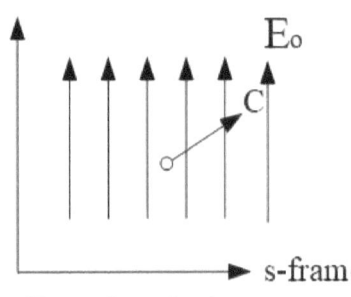

Charge's velocity,

$$C = (C_x, C_y, 0),$$

in a static **E** field

Charge's velocity,

$$C' = (0, C_{y'}, 0),$$

Flux's velocity is $-v = -C_x$

$$C = (C_x, C_y, 0) \qquad\qquad C' = (0, \gamma_x C_y, 0)$$
$$E = (0, E_o, 0) \qquad\qquad E' = (0, \gamma_x E_o, 0)$$
$$cB = (0, 0, 0) \quad \text{From case 1: } B' = (0, 0, -\gamma_x\beta_x E_o/c),$$
$$cB' = (0, 0, -\gamma_x C_x E_o)$$

$$F = (0, qE_o, 0) \text{ (Trivial for this case)}$$

Case 4 s'-frame;

$$F_{x'} = \frac{F_x - v(\vec{C}\cdot\vec{F})}{1 - vC_x} = \frac{0 - C_x(0 + C_y qE_o + 0)}{1 - C_x C_x} = \gamma_x^2 C_x C_y qE_o$$

$$F_{y'} = \frac{F_y\sqrt{1-v^2}}{1-vC_x} = \frac{qE_o\sqrt{1-C_x^2}}{1-C_x^2} = \gamma_x qE_o \;,\; F_{z'} = F_z = 0$$

$$\vec{F'} = qE_o\gamma_x\left(-\gamma_x C_x C_y, 1, 0\right) \;,\; \text{Then for EM}$$

$$\vec{F'} = q\left(\vec{E'} + \vec{v'}\times c\vec{B'}\right) \text{ with } \vec{v'} = \vec{C'}$$

A Tachyon Theory of Everything

Continuing with the cross product;

$$\vec{C} \times c\vec{B'} = \begin{vmatrix} i & j & k \\ 0 & \gamma_x C_y & 0 \\ 0 & 0 & -\gamma_x C_x E_o \end{vmatrix} = i(-\gamma_x^2 C_x C_y E_o - 0) - j(0 - 0) + k(0 - 0)$$

$$\vec{F'} = q\Big((0,\ \gamma_x E_o,\ 0) + (-\gamma_x^2 C_x C_y E_o,\ 0,\ 0)\Big) = qE_o\gamma_x(-\gamma_x C_x C_y,\ 1,\ 0)$$

Case 4 s"-frame with boost $v' = (0,\ C_{y'},\ 0)$, I permutate the indexes in the Transformation equations with x→y, y→z, and z→x, so $F_{y"}$ transforms like F_x ;

$$F_{y"} = \frac{F_{y'} - v'(\vec{C'} \cdot \vec{F'})}{1 - v'C_{y'}}$$

$$F_{y"} = \frac{qE_o\gamma_x - C_{y'}\Big((0,\ C_{y'},\ 0) \cdot (qE_o\gamma_x^2 C_x C_y,\ qE_o\gamma_x,\ 0)\Big)}{1 - C_{y'}C_{y'}}$$

$$F_{y"} = \frac{qE_o\gamma_x - C_{y'}(0 + C_{y'}qE_o\gamma_x + 0)}{1 - C_{y'}^2} = \frac{qE_o\gamma_x(1 - C_{y'}^2)}{1 - C_{y'}^2} = qE_o\gamma_x$$

$$F_{x"} = \frac{F_{x'}\sqrt{1 - v'^2}}{1 - v'C_{y'}} = \frac{-qE_o\gamma_x^2 C_x C_y\sqrt{1 - C_{y'}^2}}{1 - C_{y'}C_{y'}} \quad \text{using } \gamma_x C_y = C_{y'}$$

$$F_{x"} = -qE_o\gamma_x\gamma_{y'}C_x C_{y'} \quad \text{then with } F_{z"} = 0 \ ,$$

$$\vec{F"} = qE_o\gamma_x(-\gamma_{y'}C_x C_{y'},\ 1,\ 0)$$

106

Appendix Traditional Answers

Then for the EM solution;

$$\vec{F''} = q\left(\vec{E''} + \vec{v''} \times c\vec{B''}\right) \text{ but with } \vec{v''} = \vec{C''} = \vec{0} \ , \ \vec{v''} \times c\vec{B''} = 0$$

So, only need $\vec{E''}$

$$E_{y''} = E_{y'} = \gamma_x E_o \ , \ E_{z''} = \gamma_{y'}\left(E_{z'} - v'cB_{x'}\right) = 0$$

$$E_{x''} = \gamma_{y'}\left(E_{x'} + v'cB_{z'}\right) = \gamma_{y'}\left(0 + C_{y'}\left(-\gamma_x C_x E_o\right)\right) = -\gamma_x \gamma_{y'} C_x C_{y'} E_o$$

So $\vec{E''} = \left(-\gamma_x \gamma_{y'} C_x C_{y'} E_o, \ \gamma_x E_o, \ 0\right)$, and

$$\vec{F''} = qE_o \gamma_x \left(-\gamma_{y'} C_x C_{y'}, \ 1, \ 0\right)$$

Case 5 is the s'-frame with and arbitrary boost so both the charge and the flux have a velocity in s'-frame;

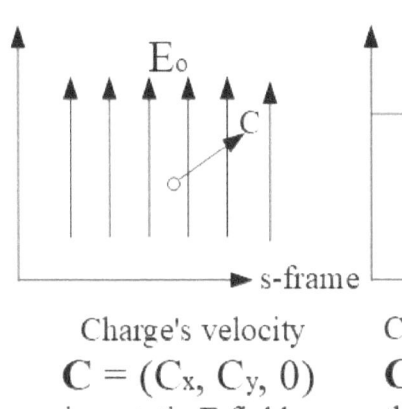
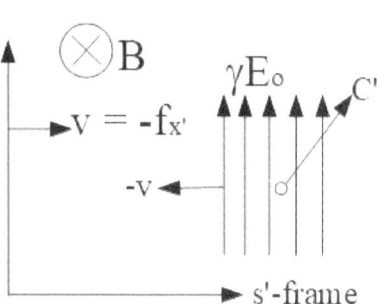

Charge's velocity

$C = (C_x, C_y, 0)$
in a static **E** field

$c\mathbf{B} = (0, 0, 0)$, and

$\mathbf{F} = (0, qE_o, 0)$

From earlier cases

Charge's velocity

$\mathbf{C'} = (C_{x'}, C_{y'}, 0)$
the flux has a velocity of

$\mathbf{f_o'} = (-v, 0, 0)$ then

$\mathbf{f'} = (-v, C_{y'}, 0)$

$$\mathbf{E'} = (0, \gamma_v E_o, 0), \quad c\mathbf{B'} = (0, 0, -v\gamma_v E_o)$$

Again, same s-frame as previous cases, so aside from the specific boost velocity, v, the s'-frame's field configuration is the same. Start with force transformation;

$$F_{x'} = \frac{F_x - v(\vec{C} \cdot \vec{F})}{1 - vC_x} = \frac{0 - (-f_{x'})((C_x, C_y, 0) \cdot (0, qE_o, 0))}{1 - (-f_{x'})C_x} = \frac{f_{x'}C_y qE_o}{1 + f_{x'}C_x}$$

Where $C_x = \dfrac{C_{x'} + (-f_{x'})}{1 + (-f_{x'})C_{x'}} = \dfrac{C_{x'} - f_{x'}}{1 - f_{x'}C_{x'}}$,and $C_y = \dfrac{C_{y'}\sqrt{1 - f_{x'}^2}}{1 - f_{x'}C_{x'}}$

So, plugging in;

$$F_{x'} = \frac{f_{x'}\left[\dfrac{C_{y'}\sqrt{1-f_{x'}^2}}{1-f_{x'}C_{x'}}\right]qE_o}{1+f_{x'}\left(\dfrac{C_{x'}-f_{x'}}{1-f_{x'}C_{x'}}\right)} = \frac{f_{x'}C_{y'}\sqrt{1-f_{x'}^2}\,qE_o}{1-f_{x'}C_{x'}+f_{x'}C_{x'}-f_{x'}^2} = \frac{qE_of_{x'}C_{y'}}{\sqrt{1-f_{x'}^2}}$$

Then for the y − component:

$$F_{y'} = \frac{F_y\sqrt{1-f_{x'}^2}}{1-(-f_{x'})C_x} = \frac{qE_o\sqrt{1-f_{x'}^2}}{1+f_{x'}\left(\dfrac{C_{x'}-f_{x'}}{1-f_{x'}C_{x'}}\right)} = \frac{qE_o(1-f_{x'}C_{x'})}{\sqrt{1-f_{x'}^2}}$$

For an answer of;

$$\vec{F'} = \left[\frac{qE_of_{x'}C_{y'}}{\sqrt{1-f_{x'}^2}}, \frac{qE_o(1-f_{x'}C_{x'})}{\sqrt{1-f_{x'}^2}}, 0\right] = \frac{qE_o}{\sqrt{1-f_{x'}^2}}(f_{x'}C_{y'}, 1-f_{x'}C_{x'}, 0)$$

Note that since the flux's velocity, $f_x = -v$, the frame's velocity, the "more traditional" answer looks like $qE_o\gamma(-vC_{y'}, 1+vC_{x'}, 0)$. I'm using the flux velocity since that is a physical thing (as opposed to just space.) that is moving at speed v in the s'-frame.

Continuing with **E** and **B** fields;

$$E_{x'} = E_x = 0 \;,\; E_{y'} = \gamma_v(E_y - vcB_z) = \frac{E_o - (-f_{x'})0}{\sqrt{1-f_{x'}^2}} = \frac{E_o}{\sqrt{1-f_{x'}^2}} \text{ and}$$

$$E_z = B_y = 0 \text{ so, } \vec{E'} = \left(0, \frac{E_o}{\sqrt{1-f_{x'}^2}}, 0\right)$$

Then with $B_{x'} = B_{y'} = 0$ and $cB_{z'} = \gamma_v(cB_z - vE_y) = \dfrac{-(-f_{x'})E_o}{\sqrt{1-f_{x'}^2}}$

$c\vec{B}' = \left(0, 0, \dfrac{f_{x'}E_o}{\sqrt{1-f_{x'}^2}}\right)$ so, $\vec{C}' \times c\vec{B}' = \begin{vmatrix} i & j & k \\ C_{x'} & C_{y'} & 0 \\ 0 & 0 & \dfrac{f_{x'}E_o}{\sqrt{1-f_{x'}^2}} \end{vmatrix}$

$\vec{C}' \times c\vec{B}' = i\left(\dfrac{C_{y'}f_{x'}E_o}{\sqrt{1-f_{x'}^2}} - 0\right) - j\left(\dfrac{C_{x'}f_{x'}E_o}{\sqrt{1-f_{x'}^2}} - 0\right) + k(0 - 0)$

$\vec{F}' = q\left[\left(0, \dfrac{E_o}{\sqrt{1-f_{x'}^2}}, 0\right) + \left(\dfrac{C_{y'}f_{x'}E_o}{\sqrt{1-f_{x'}^2}}, \dfrac{-C_{x'}f_{x'}E_o}{\sqrt{1-f_{x'}^2}}, 0\right)\right]$

$\vec{F}' = \dfrac{qE_o}{\sqrt{1-f_{x'}^2}}(C_{y'}f_{x'}, 1 - C_{x'}f_{x'}, 0)$

Again, with $f_{x'} = -v$, the "more traditional" answer is

$\mathbf{F'} = qE_o\gamma(-vC_{y'}, 1+vC_{x'}, 0)$

Case 6: Take Case 5's s'-frame and boost with $v' = (0, 0, -C_{z''})$. The page is going away from you along the depicted **B** field. Permutate the components of the transformation equations with; $x \to z$, $z \to y$, and $y \to x$

Starting with transforming the force from the Case 5 s'-frame,

$$\vec{F'} = \frac{qE_o}{\sqrt{1-f_{x'}^2}}(f_{x'}C_{y'}, \ 1-f_{x'}C_{x'}, \ 0) \ \text{Now } F_z \text{ tranforms like it's } F_x$$

$$F_{z''} = \frac{F_{z'} - v'\left(\vec{C'} \cdot \vec{F'}\right)}{1 - C_{z'}v'} \ , \text{ subbing in the s' - variables;}$$

$$F_{z''} = \frac{0 - (-C_{z''})\left((C_{x'}, C_{y'}, 0) \cdot \left(\frac{qE_o}{\sqrt{1-f_{x'}^2}}(f_{x'}C_{y'}, \ 1-f_{x'}C_{x'}, \ 0)\right)\right)}{1 - 0v'}$$

$$F_{z''} = \frac{qE_oC_{y'}C_{z''}}{\sqrt{1-f_{x'}^2}} \ . \text{ To put it all in the s'' - frame sub in,}$$

$$C_{y'} = \frac{C_{y''}\sqrt{1-(v')^2}}{1+C_{z''}v} = \frac{C_{y''}\sqrt{1-C_{z''}^2}}{1+C_{z''}(-C_{z''})} = \frac{C_{y''}}{\sqrt{1-C_{z''}^2}}$$

Similarly, $f_{x'} = \dfrac{f_{x''}}{\sqrt{1-C_{z''}^2}}$, and for later $C_{x'} = \dfrac{C_{x''}}{\sqrt{1-C_{z''}^2}}$;

to get;

$$F_{z''} = \frac{qE_o\left(\dfrac{C_{y''}}{\sqrt{1-C_{z''}^2}}\right)C_{z''}}{\sqrt{1-\left(\dfrac{f_{x''}}{\sqrt{1-C_{z''}^2}}\right)^2}} = \frac{qE_oC_{y''}C_{z''}}{\sqrt{1-f_{x''}^2-C_{z''}^2}}$$

Continuing with the x and y-components;

$$F_{x''} = \frac{F_{x'}\sqrt{1-(v')^2}}{1-C_{z'}v'} = \frac{\left(\dfrac{qE_o f_{x'} C_{y'}}{\sqrt{1-f_{x'}^2}}\right)\sqrt{1-C_{z''}^2}}{1-0(-C_{z''})} = \frac{qE_o f_{x'} C_{y'}\sqrt{1-C_{z''}^2}}{\sqrt{1-f_{x'}^2}}$$

and for s'';

$$F_{x''} = \frac{qE_o\left(\dfrac{f_{x''}}{\sqrt{1-C_{z''}^2}}\right)\left(\dfrac{C_{y''}}{\sqrt{1-C_{z''}^2}}\right)\sqrt{1-C_{z''}^2}}{\sqrt{1-\left(\dfrac{f_{x''}}{\sqrt{1-C_{z''}^2}}\right)^2}} = \frac{qE_o f_{x''} C_{y''}}{\sqrt{1-f_{x''}^2-C_{z''}^2}}$$

Onto y'' ;

$$F_{y''} = \frac{F_{y'}\sqrt{1-(v')^2}}{1-C_{z'}v'} = \frac{\left|\dfrac{qE_o(1-f_{x'}C_{x'})}{\sqrt{1-f_{x'}^2}}\right|\sqrt{1-C_{z''}^2}}{1-0(-C_{z''})}$$

$$F_{y''} = \frac{qE_o(1-f_{x'}C_{x'})\sqrt{1-C_{z''}^2}}{\sqrt{1-f_{x'}^2}}$$

and for s'';

$$F_{y''} = \frac{qE_o\left(1-\left(\dfrac{f_{x''}}{\sqrt{1-C_{z''}^2}}\right)\left(\dfrac{C_{x''}}{\sqrt{1-C_{z''}^2}}\right)\right)\sqrt{1-C_{z''}^2}}{\sqrt{1-\left(\dfrac{f_{x''}}{\sqrt{1-C_{z''}^2}}\right)^2}}$$

$$F_{y''} = \frac{qE_o(1-f_{x''}C_{x''}-C_{z''}^2)}{\sqrt{1-f_{x''}^2-C_{z''}^2}}$$ Putting it all together:

$$\vec{F''} = \frac{qE_o}{\sqrt{1-f_{x''}^2-C_{z''}^2}}(f_{x''}C_{y''},\ 1-f_{x''}C_{x''}-C_{z''}^2,\ C_{y''}C_{z''})$$

Appendix Traditional Answers

Continuing with the EM solution, the **E'** and **B'** fields with $v = -f_{x'}$;

$$\vec{E'} = \left(0, \frac{E_o}{\sqrt{1-f_{x'}^2}}, 0\right) \text{ and } c\vec{B'} = \left(0, 0, \frac{f_{x'}E_o}{\sqrt{1-f_{x'}^2}}\right)$$

$$E_{z''} = E_{z'} = 0 \;,\; E_{y''} = \gamma_{v'}\left(E_{y'} + v'cB_{x'}\right) = \frac{1}{\sqrt{1-C_{z''}^2}}\frac{E_o}{\sqrt{1-f_{x'}^2}} \;,\; \text{and}$$

$$E_{x''} = \gamma_{v'}\left(E_{x'} - v'cB_{y'}\right) = 0 \;,\text{so}\; \vec{E''} = \left(0, \frac{1}{\sqrt{1-C_{z''}^2}}\frac{E_o}{\sqrt{1-f_{x'}^2}}, 0\right)$$

Then $cB_{z''} = cB_{z'} = \dfrac{f_{x'}E_o}{\sqrt{1-f_{x'}^2}}$, $cB_{y''} = \gamma_{v'}\left(cB_{y'} - v'E_{x'}\right) = 0$,and

$$cB_{x''} = \gamma_{v'}\left(cB_{x'} + v'E_{y'}\right) = \frac{1}{\sqrt{1-C_{z''}^2}}\left(0 + (-C_{z''})\frac{E_o}{\sqrt{1-f_{x'}^2}}\right)$$

$$cB_{x''} = \frac{-C_{z''}E_o}{\sqrt{1-C_{z''}^2}\sqrt{1-f_{x'}^2}} \;,\text{so}$$

$$c\vec{B''} = \left(\frac{-C_{z''}E_o}{\sqrt{1-C_{z''}^2}\sqrt{1-f_{x'}^2}}, 0, \frac{f_{x'}E_o}{\sqrt{1-f_{x'}^2}}\right)$$

So the cross product in the Lorentz force is;

$$\vec{C''} \times c\vec{B''} = \begin{vmatrix} i & j & k \\ C_{x''} & C_{y''} & C_{z''} \\ \dfrac{-C_{z''}E_o}{\sqrt{1-C_{z''}^2}\sqrt{1-f_{x'}^2}} & 0 & \dfrac{f_{x'}E_o}{\sqrt{1-f_{x'}^2}} \end{vmatrix}$$

$$\vec{C''} \times c\vec{B''} = i\left(\frac{C_{y''}f_{x'}E_o}{\sqrt{1-f_{x'}^2}} - 0 \right) - j\left(\frac{C_{x''}f_{x'}E_o}{\sqrt{1-f_{x'}^2}} - \frac{-C_{z''}^2 E_o}{\sqrt{1-C_{z''}^2}\sqrt{1-f_{x'}^2}} \right)$$

$$+ k\left(0 - \frac{-C_{y''}C_{z''}E_o}{\sqrt{1-C_{z''}^2}\sqrt{1-f_{x'}^2}} \right)$$

$$\vec{C''} \times c\vec{B''} = \frac{E_o}{\sqrt{1-f_{x'}^2}}\left(\frac{C_{y''}f_{x'}}{} , \frac{-C_{z''}^2}{\sqrt{1-C_{z''}^2}} - \frac{C_{x''}f_{x'}}{} , \frac{C_{y''}C_{z''}}{\sqrt{1-C_{z''}^2}} \right)$$

Plugging into the Lorentz Force;

$$\vec{F''} = q\left(0, \frac{E_o}{\sqrt{1-f_{x'}^2}\sqrt{1-C_{z''}^2}}, 0 \right)$$

$$+ q\frac{E_o}{\sqrt{1-f_{x'}^2}}\left(C_{y''}f_{x'} , \frac{-C_{z''}^2}{\sqrt{1-C_{z''}^2}} - C_{x''}f_{x'} , \frac{C_{y''}C_{z''}}{\sqrt{1-C_{z''}^2}} \right)$$

$$\vec{F''} = \frac{qE_o}{\sqrt{1-f_{x'}^2}}\left(C_{y''}f_{x'} , \frac{1-C_{z''}^2}{\sqrt{1-C_{z''}^2}} - C_{x''}f_{x'} , \frac{C_{y''}C_{z''}}{\sqrt{1-C_{z''}^2}} \right)$$

Substituting in the transforms for the s"-frame;

$$\vec{F^n} = \cfrac{qE_o}{\sqrt{1 - \left(\cfrac{f_{x''}}{\sqrt{1 - C_{z''}^2}}\right)^2}}$$

$$\times \left(C_{y''}\left(\frac{f_{x''}}{\sqrt{1 - C_{z''}^2}}\right), \frac{1 - C_{z''}^2}{\sqrt{1 - C_{z''}^2}} - C_{x''}\left(\frac{f_{x''}}{\sqrt{1 - C_{z''}^2}}\right), \frac{C_{y''}C_{z''}}{\sqrt{1 - C_{z''}^2}}\right)$$

Finally;

$$\vec{F^n} = \frac{qE_o}{\sqrt{1 - C_{z''}^2 - f_{x''}^2}}\left(C_{y''}f_{x''}, \ 1 - f_{x''}C_{x''} - C_{z''}^2, \ C_{y''}C_{z''}\right)$$

My purpose for including these traditional solutions was one-I did them to check my own work, so having done them I may as well include them, and two- to point out how doing the Force Transforms allows me to skip considering the **B** field since in the rest frame of the charge, only the **E** field is responded to, and that force can be transformed to the frame with a magnetic response.

Bibliography

1) A First Course in General Relativity, by Bernard F Schutz, published by Cambridge University Press. 1990

2) Tables of Integrals,Series,and Products Corrected and Enlarged Edition, by I.S. Gradshteyn and I.M Ryzhik Published by Academic Press 1980

3) Introduction to Special Relativity, by Robert Resnick, published by John Wiley & Sons 1968

4) Differential Geometry, by Erwin Kreyszig, published by Dover Publications 1991

5) All of my undergraduate math and physics texts.

6) Wikipedia was helpful for quickly looking up a formula or two.

7) Andres J. Kalnay, "Complex Physical Quantities and space like states"
Tachyons, Monopoles, and Related Topics, Erasmo Recami, North Holland Publishing.

www.ingramcontent.com/pod-product-compliance
Lightning Source LLC
Chambersburg PA
CBHW030657220526
45463CB00005B/1819